GLOBAL AGRICULTURE: DEVELOPMENTS, ISSUES, AND RESEARCH

GLOBAL AGRICULTURE

DEVELOPMENTS, ISSUES, AND RESEARCH

VOLUME 1

Global Agriculture: Developments, Issues, and Research

Additional books in this series can be found on Nova's website under the Series tab.

Additional e-books in this series can be found on Nova's website under the e-books tab.

GLOBAL AGRICULTURE: DEVELOPMENTS, ISSUES, AND RESEARCH

GLOBAL AGRICULTURE

DEVELOPMENTS, ISSUES, AND RESEARCH

VOLUME 1

MARVIN R. ROBERTSON
AND
JEFFERY I. SCHMIDT
EDITORS

nova publishers
New York

For permission to use material from this book please contact us:
Telephone 631-231-7269; Fax 631-231-8175
Web Site: http://www.novapublishers.com

LIBRARY OF CONGRESS CATALOGING-IN-PUBLICATION DATA

ISBN: 978-1-62257-369-1

ISSN: 2168-0434

Published by Nova Science Publishers, Inc. † New York

CONTENTS

Preface		**vii**
Chapter 1	Peru: An Emerging Exporter of Fruits and Vegetables *Birgit Meade, Katherine Baldwin and Linda Calvin*	**1**
Chapter 2	The U.S. and Mexican Dry Bean Sectors *Steven Zahniser, Marcela Vera Torres, José Alberto Cuéllar Álvarez, Nicolás Fernando López López and Rhea Bhatta*	**39**
Chapter 3	Investment in Processing Industry Turns Chinese Apples into Juice Exports *Fred Gale, Sophia Huang and Yingying Gu*	**87**
Chapter 4	Trade and Food Security Implications from the Indonesian Agricultural Experience *Nicholas Rada and Anita Regmi*	**117**
Index		**137**

PREFACE

This book explores the developments, issues and research outlook of global agriculture. Topics discussed include the trade and food security implications from the Indonesian agricultural experience; the remarkable growth in China's apple juice concentrate exports since the 1990s; the U.S. and Mexican dry bean sectors; and Peru as an emerging exporter of fruits and vegetables.

Chapter 1 – Over the past two decades, Peru has emerged as a significant fruit and vegetable exporter. The United States is one of its most important markets. This rise in exports was driven by advantages in climate, low production costs, and preferential trading arrangements with major countries. But because many of the same products exported by Peru are grown in the United States, increased Peruvian exports will continue to affect U.S. growers as well as consumers. In this report, we highlight the success stories of three commodities—asparagus, processed artichokes, and table grapes—that represent different impacts on U.S. growers. Despite Peru's success in developing a fruit and vegetable export industry, challenges remain. In order to continue on its current export trajectory, the country must confront a number of constraints, including water access, land rights, and underdeveloped export infrastructure.

Chapter 2 – Liberalization of U.S.-Mexico dry bean trade, as part of the North American Free Trade Agreement (NAFTA), has enabled U.S. dry bean producers to provide a steadier and larger portion of Mexican supply. Long-term prospects for the newly integrated market are shaped by various factors. On the supply side, the future structure of Mexico's dry bean sector will be comprised of a smaller number of producers who work larger plots of land for higher yields. U.S. dry bean production is already characterized by large-scale farms, and recent yields exhibit a modest upward trend. On the demand side, per capita disappearance of dry beans in Mexico is declining over the long term as consumers shift away from traditional staple foods. U.S. per capita consumption is relatively stable but only at about a fourth of Mexico's average level, even though dry beans are a low-fat source of important nutrients.

Chapter 3 – In this report, ERS profiles the remarkable growth in China's apple juice concentrate exports since the 1990s and analyzes the factors behind the boom. Investment in the juice processing industry brought apples from China to the world market in the form of apple juice concentrate. The United States is the largest buyer, and concentrate from China now accounts for two-thirds of the U.S. apple juice supply. China's juice industry sustained its dramatic growth by expanding into the country's hinterland with support from officials eager to develop the poor northwestern region. In recent years, rising apple prices have begun to act as a brake on the industry's expansion, and the industry faces challenges in improving the quality of apples used for juice processing.

Chapter 4 – As the Indonesian Government transitioned from a food (rice) self-sufficiency focus to an industrial export-oriented development strategy in the mid-1980s, the agricultural sector responded by diversifying production to include cash crops. Sharp currency devaluation and trade liberalization following the Asian financial crisis in 1997 provided incentives for Indonesian agricultural producers to assert their comparative advantage in tropical perennial crops. This report evaluates the ensuing impact on Indonesia's agricultural productivity growth and consumer food demand and lifestyle. Shifting agricultural production and consumption patterns have led to improving agricultural trade patterns and food security. These developments have contributed to increased export opportunities for U.S. agricultural suppliers.

In: Global Agriculture. Volume 1
Editors: M.R. Robertson, J.I. Schmidt ISBN: 978-1-62257-369-1

Chapter 1

PERU: AN EMERGING EXPORTER OF FRUITS AND VEGETABLES*

Birgit Meade, Katherine Baldwin and Linda Calvin

ABSTRACT

Over the past two decades, Peru has emerged as a significant fruit and vegetable exporter. The United States is one of its most important markets. This rise in exports was driven by advantages in climate, low production costs, and preferential trading arrangements with major countries. But because many of the same products exported by Peru are grown in the United States, increased Peruvian exports will continue to affect U.S. growers as well as consumers. In this report, we highlight the success stories of three commodities—asparagus, processed artichokes, and table grapes—that represent different impacts on U.S. growers. Despite Peru's success in developing a fruit and vegetable export industry, challenges remain. In order to continue on its current export trajectory, the country must confront a number of constraints, including water access, land rights, and underdeveloped export infrastructure.

---

* This is an edited, reformatted and augmented version of the United States Department of Agriculture publication, FTS-345-01, A Report from the Economic Research Service from www.ers.usda.gov, dated December 2010.

Keywords: Peru, fruit and vegetable exports, asparagus, artichokes, table grapes, California, Michigan, Washington, water management, land rights, infrastructure

Introduction

Since the mid-1980s, Peru has emerged as an important fruit and vegetable exporter, and the United States is one of the country's top export destinations. The question is whether or not Peru will be able to sustain and expand its role as an important player in the international fruit and vegetable trade. Traditionally, Peru has been known mostly for its exports of metals and ores, but since the early 1990s, Peru's horticultural exports have grown at an annual rate of 16 percent, outpacing any other commodity group (Food and Agriculture Organization of the United Nations (FAO), various years). The early 1990s were a time when a new constitution and new laws encouraged Peruvian engagement in international trade and investment in Peru. These developments, along with Peru's natural advantages, such as a climate ideally suited for production of many horticultural crops, provided the groundwork for ongoing strong economic and export growth in traditional, as well as new, fruit and vegetable products.

The United States also produces many of the fruits and vegetables that it imports from Peru. A second question, therefore, is what impact Peru's increasing importance as a U.S. supplier and competitor will have on U.S. production, trade, and consumption. Although Peru's export boom began with asparagus, the country's export portfolio has since expanded to include such diverse products as processed artichokes, avocados, bananas, various types of citrus fruit, grapes, mangoes, onions, and paprika.

This report presents Peru as an agricultural producer and trading partner, with a special focus on the fruit and vegetable export industry. We highlight the factors that make Peru a successful fruit and vegetable exporter. Peru's main advantages are its (1) investor-friendly business environment and policy framework, (2) free or preferential trade agreements with major importers, such as the United States, the European Union (EU), and China, (3) relatively cheap labor, and (4) good climate.

We then focus on three in-depth commodity analyses of Peruvian fruit and vegetable industries that can be described as success stories, but with differing levels of impact on U.S. growers. Asparagus is an example of a commodity in which Peru competes directly with U.S. producers. Artichokes are another successful export commodity; Peruvian exports to the United States are almost exclusively processed artichokes. U.S. artichoke growers have not produced for the processed market since 2000, so Peru does not compete with U.S. production. This type of trade relationship is also true for a number of other Peruvian export products, such as mangoes (the United States produces very few mangoes) and piquillo peppers. Our third case study concentrates on table grapes. Because of the timing of the Peruvian growing season, exports to the United States compete with U.S.-produced grapes in the U.S. market for only a very short time and Peruvian shipments are very small compared with the U.S. supply. However, grape exports from both countries do compete in third-country markets, which is a concern to U.S. growers.

We point out some challenges to Peru, such as its reliance on a small number of export products and markets, the large distance to its export markets, shortcomings in its export infrastructure, water scarcity in the coastal areas, and unresolved land right issues. We conclude with an assessment of Peru's potential to continue expanding production and trade in the medium term and likely impacts on U.S. production, trade, and consumption.

PERU: AN INCREASINGLY IMPORTANT EXPORTER OF FRUITS AND VEGETABLES

Peru is a country of 29 million people situated on the Pacific coast of South America. It is classified as a lower middle-income country by the World Bank and had a per capita gross national income (GNI) of $3,990 in 2008. Gross domestic product (GDP) grew at 9.8 percent in 2008 (World Bank, 2010), continuing the recent trend of strong economic growth.[1] Exports were an important driver of this growth, with merchandise exports of $31.5 billion in 2008.

Exports of goods and services accounted for 27.1 percent of GDP in 2008, up from 25.6 percent in 2007 and only 12.6 percent in 1991. A major part of these exports consists of mineral commodities, which benefited significantly from rising international prices through 2007.[2]

Agricultural exports accounted for 7.2 percent of all merchandise trade in 2007 (FAO, various years).

Peru has also been successful at attracting foreign direct investment (FDI)—U.S. $5.3 billion in 2007, up from $3.5 billion in 2006—an important source of capital, which, so far, has been mostly going into the petroleum/

mining sector. FDI contributes to expanding the export sector and investments in major infrastructure projects (World Bank, 2010).

While coffee remains Peru's most important agricultural export crop, more than 60 percent of all agricultural exports are now fruits and vegetables. Asparagus is the largest fruit and vegetable export item (fig. 1). Peru’s fruit and vegetable exports were close to U.S. $1.2 billion in 2009 (Global Trade Information Services, Inc. (GTIS), 2010), up from $60 million in 1990 (FAO, 2009).

Since 1990, exports have increased at an average annual rate of 16 percent—a faster growth rate than Peruvian merchandise exports as a whole (FAO, 2009).

By comparison, the U.S. share of fruit and vegetable exports was close to half a billion dollars in 2009.

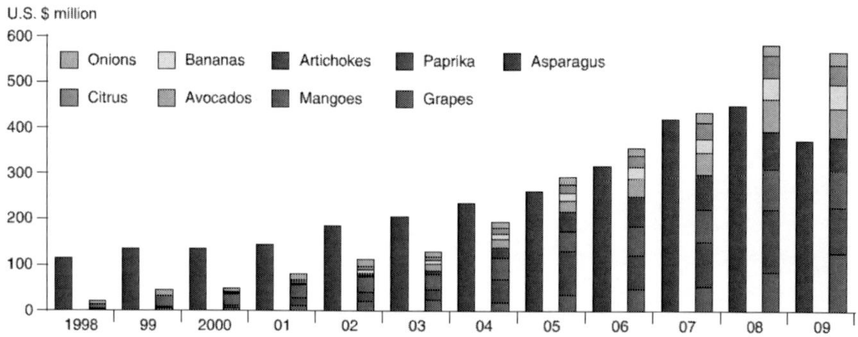

Source: Global Trade Information Services, Inc. (2010).

Figure 1. Growth in the value of selected Peruvian fruit and vegetable exports, 1998-2009.

In general, the United States and Europe are Peru’s largest export markets (table 1). In 2009, the United States accounted for 40 percent of Peru’s fruit and vegetable exports and Europe accounted for 44 percent (GTIS, 2010). In 2009, over half of all Peruvian exports of fresh asparagus, canned artichokes, and fresh onions were shipped to the United States.

Recently, China has emerged as an important export market for Peruvian grapes. For fresh products, it will become increasingly important for Peru to find market niches where it can be the low-cost supplier in a particular seasonal window, as competition from lower cost producers is increasing. For the processed market, seasons are not as important, but keeping costs low is critical.

Table 1. Selected Peruvian fruit and vegetable exports, 2009

		Share shipped to selected destinations			
Product	Value	United States	EU	Northeast Asia	Rest of world
	U.S. $ million			*Percent*	
Asparagus	373.9	50.2	44.1	1.3	4.4
Grapes	126.6	27.8	31.6	18.9	21.7
Paprika	101.3	37.8	39.1	0.1	23.0
Mango	81.2	29.9	62.1	1.4	6.6
Artichoke	72.3	59.7	36.8	0.0	3.5
Avocado	64.1	0.2	95.2	0.0	4.6
Bananas	51.6	24.3	63.4	11.7	0.5
Citrus	43.4	26.7	52.5	0.0	20.7
Onion	28.1	77.4	4.0	0.0	18.5

EU = European Union.

Source: Global Trade Information Services, Inc. (2010).

WHAT FACTORS EXPLAIN PERU'S SUCCESS IN EXPORTS OF FRUITS AND VEGETABLES?

Peru's fruit and vegetable export industry benefits from a combination of factors that have supported its rapid growth. In the early 1990s, a new constitution and new laws encouraged international investment in Peru, as well as Peruvian engagement in international trade. Peru's export industry benefits from a business climate that welcomes foreign investors and encourages trade. Free trade agreements with all major trading partners, such as the United States, the EU, and China, provide an invaluable stimulus to the export industry. The agricultural sector continues to be very labor intensive, and Peru's fruit and vegetable industry has been fortunate to have had access to abundant and relatively cheap farm labor. Lastly, Peru's climate is very well suited for the production of a range of fruits and vegetables.

Business Environment

Peru's economy stabilized after the Latin American debt crisis in the 1980s. During the 1990s, inflation was kept below 5 percent thanks to structural reforms, deregulation, fiscal restraint, and international loans (The Economist Intelligence Unit, 1997 and 2008). Also, the Government of Peru created a business and investment-friendly environment with its 1991 Foreign Investment Promotion Law, which legalized unrestricted private foreign land ownership and investment and outlawed discrimination between foreign and domestic investors. This law and the Government's campaign of privatization in the fishing, mining, telecommunication, energy, and finance sectors were successful in attracting foreign investors. In 2007, Peru's net FDI was close to U.S. $5.3 billion (World Bank, 2009b).

This favorable business environment and relaxed land ownership rights have supported capable producers and agricultural entrepreneurs who have adopted state-of-the-art technology and organizational standards. Large-scale exporters are fully aware of international market expectations. Many of them have founded or joined associations that can effectively promote their commodity, as the associations have a keen understanding of their respective industries, as well as world market conditions.[3]

Trade Agreements

This investor-friendly business climate has contributed to the signing of several trade agreements. The United States and Peru have a long-established preferential trading relationship, going back to the 1991 Andean Trade Preference Act (ATPA). The ATPA was a multilateral agreement between the United States and Bolivia, Colombia, Ecuador, and Peru that offered these Andean nations trade concessions in order to establish legitimate industries and to reduce dependence on drug production and trafficking. These concessions included preferential access to the U.S. market for a large share of their exports, including duty-free access for most fruit and vegetable products. 4

However, the ATPA was scheduled to expire in 2006, along with it its preferential access provisions. After annual renewals, the U.S.-Peru Trade Promotion Agreement (PTPA) replaced the earlier, less far-reaching trade agreement in February 2009. The PTPA is a comprehensive free trade agreement that makes Peru's preferential access to the U.S. market permanent. The agreement also immediately eliminated most of Peru's tariffs on U.S. exports, with all remaining tariffs phased out over defined periods (within less than 5 or 10 years, but no more than 17 years) (U.S. Trade Representative, 2009). The agreement furthermore addressed several nontariff issues, such as individual property rights and dispute settlement, and thus eliminated a range of trade-inhibiting obstacles (Villarreal, 2006). Having preferential access to the U.S. market is of great value to the Peruvian export sector, even if some barriers arising from sanitary and phytosanitary problems remain.

Peru is working toward broadening its trade base and has signed free trade agreements (FTAs) with most major trading partners, most notably with China and the EU (Andina, 2010). FTAs with Japan and South Korea are being negotiated in 2010, and further FTAs are planned with Russia, India, Morocco, and South Africa.

Labor

Relatively low labor costs have helped Peru to compete successfully in the world market. As a lower middle-income country with a very skewed income distribution, average wages are far below those of many of its competitors. Low labor costs are partly a consequence of high *under*employment (less working hours than desired), estimated at 64.5 percent in 2005 (Bolle and Villareal, 2007) even though the official *un*employment rate has been around 8 percent in recent years. In 2008, the average nominal wage was only about $388 per month (The Economist Intelligence Unit, 2010), or roughly $2.15 per hour.[5] Agricultural labor costs are much lower than this average. Low agricultural wages are a result of an abundant supply of unskilled rural workers. However, workers employed by agricultural exporting companies earn about 30 percent more than those working for agricultural companies that serve the local market. Growers in Ica reported that labor costs were about $8 per worker per day in 2008, higher than in other parts of the country. Ica has reached the point where labor for the exporting agricultural companies is getting scarce, driving wages higher. Fruit and vegetable production continues to be labor intensive, as many tasks, such as the harvest of asparagus, are not mechanized.[6]

Although labor costs remain low compared with those in competitor countries, jobs in the sector are coveted as they often come with benefits, especially in areas where suitable labor is becoming scarce, such as in Ica. Here, some growers invest in family services, such as health care units, nurseries, and other community services, to retain their workers and thus reduce costly turnover and to keep their employees personally invested in the success of the company. Peru's low labor costs help compensate for high transportation costs to major export markets and thus keep Peru competitive in the world market. Impoverished families living in the highlands (Sierra) often have no other choice than to move to the coastal area for employment even though some programs are underway to provide farmers with support to develop highland-based export industries such as avocado production.

Climate

Peru's natural advantage in fruit and vegetable production is its climate. The Peruvian fruit and vegetable export industry is located along the long, narrow coastal plain, which has a maximum width of 75 miles. Peru consists of three very distinct topographic regions that run from north to south (fig. 2): a narrow coastal desert plain along the Pacific Ocean in the west, the Andes in the center (known as the Sierra), and the jungle in the east. The Sierra and the jungle cover about three-fifths of the Peruvian land area, but these regions are sparsely populated. With such variation in climate, Peruvian growers have experimented in alternative growing regions (further north or south on the long coastal plain or up in altitude in the Sierra) in order to increase the export seasons for some commodities.

The coastal plain is a tropical desert. The temperature is relatively constant and mild due to the Humboldt Current, which brings cold water from the Antarctic to most of Peru's coast and thus moderates otherwise hot temperatures. The closeness of Peru to the equator gives the country a relatively even length of daylight throughout the year and guarantees the absence of frost. In this spring-like climate, asparagus, for example, can be harvested 12 months out of the year if proper water-management techniques are used. As a Southern Hemisphere fruit and vegetable exporter, Peru has the advantage of being able to supply the high-income and populous Northern Hemisphere markets in their off-season.

The Peruvian coastal region has virtually no rainfall (except during the El Niño phenomenon), and export agriculture depends on irrigation from wells or rivers, which can be problematic.[7] Peruvian exporters claim that, while others (for example, China) may have cheaper production costs for some crops, Peru has the most consistent supply and quality of fruits and vegetables due to its predictable weather conditions. The desert conditions are excellent for growing fruits and vegetables as long as irrigation water is adequate. Recently, however, Peruvian growers have become concerned about the possible effects of climate change, which might threaten the predictability of weather patterns since more extreme weather conditions, such as unusually high and low temperatures and flash floods, have been observed. Peruvian scientists have also expressed concern about the retreat of Andean glaciers, the source of much of the water used in coastal agriculture (Flores, 2007).

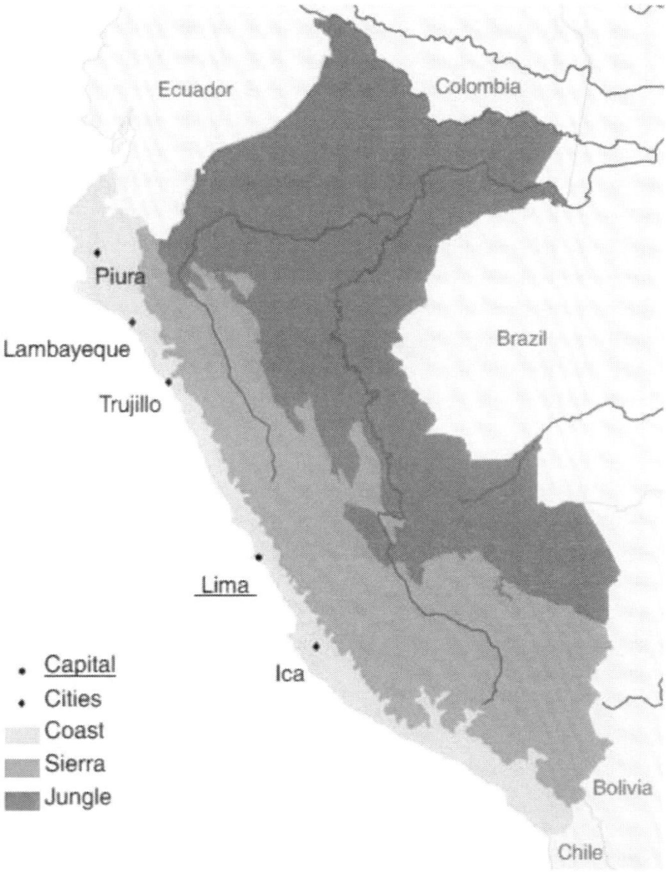

Source: USDA, Economic Research Service.

Figure 2. Map of Peru.

PERUVIAN EXPORT SUCCESS STORIES

Peru's combination of business climate, trade preferences, low labor costs, and climatic conditions helped lay the foundation for developing a competitive and successful agricultural export industry. Many of the fruit and vegetable industries grew out of seasonal demand from Northern Hemisphere trading partners. The conditions and developments described in this report determine, to a large extent, the current success and future uncertainty of Peru's export commodities.

Although each of these commodities has a different story, they also share common characteristics and trends. In examining the successful export product stories of Peru and their effect on the U.S. market, we identified three different groups of products based on their effect on their U.S. counterpart industries. The first group negatively affects the U.S. counterpart industry by directly replacing the U.S.-produced commodity within the U.S. market. Asparagus is the most prominent example, but Peruvian-produced paprika has also negatively affected U.S. production. The second group benefits U.S. consumers while having little or no impact on U.S. producers, primarily due to negligible production of those particular commodities in the United States. Although tropical fruit products, such as mangoes and bananas, also fit this category, we explore in some detail the high-value processed artichoke market.

The last group represents products that are produced both in the United States and Peru but do not compete head to head throughout their seasons, although there is some overlap of seasons, which is the case for table grapes and sweet onions.

We examine table grapes in detail. Many U.S. grape growers are concerned about competition between the United States and Peru in important third-country markets. These case studies of asparagus, processed artichokes, and table grapes, therefore, serve as examples for a number of other successful export commodities and for the potential repercussions of the ongoing fast expansion of Peruvian fruit and vegetable exports, not only for Peru, but for U.S. producers and consumers.

Success Story #1: Asparagus

Asparagus was the first, and remains the most prominent, example of Peru's success as a major exporter of fruit and vegetable products. While the emergence of Peru as a year-round global asparagus supplier was good news for U.S. consumers, it had a considerable negative effect on the U.S. asparagus industry.

The huge success of this export product is partly due to the variety of forms in which it can be produced and processed.

Production

Peru is the world's second largest asparagus producer after China. The Peruvian asparagus industry started in the early 1950s in La Libertad, the region around Trujillo, as a canned white asparagus export industry. In the mid-1980s, the U.S. Agency for International Development began working with growers in Ica to find nontraditional agricultural exports that could reduce the country's reliance on coca production. Asparagus looked promising because of the area's deep sandy soil and advantageous climatic conditions. The first commercial exports of fresh green asparagus began in the mid-1980s. Today, most production is concentrated in the departments of Ica and La Libertad, with 39 percent and 44 percent of the total asparagus production area, respectively; Lima (9 percent) and Ancash (8 percent) account for most of the remaining land in asparagus production (Ministerio de Agricultura del Perú (MINAG), 2010). While white asparagus[8] once dominated Peruvian production, in 2008 green asparagus production made up 87 percent of the total (Benson, 2010).

Total planted acreage has nearly quadrupled since 1988, with production area over 29,800 hectares in 2008 (MINAG, 2010) (fig. 3). Peru is the only country in the world that harvests asparagus year round, with two or sometimes three harvests per year per field. The large-scale growers tend to run farms with state-of-the-art technology, such as drip irrigation. They can manipulate the harvest seasons by withholding water to the plants, which enables Peru to target the most profitable market periods. Peru has the world's highest asparagus yields, increasing steadily and almost doubling since 1990 (MINAG, 2010).[9]

Exports

After coffee, asparagus is Peru's largest agricultural export product at $374 million in 2009. Exports of fresh asparagus in that year were valued at $240.6 million; exports of processed asparagus (such as asparagus in cans or jars but not frozen) added another $109.7 million in value, and frozen asparagus exports totaled $23.5 million. Domestic demand is very low for asparagus, which is not a native vegetable and has not made inroads into the Peruvian cuisine; thus, nearly all of the country's production is exported. A major share of these exports—50.2 percent, or $187.8 million, in 2009—was destined for the United States.

The share of asparagus going to the three different markets—fresh, processed, and frozen—has changed over time. These shifts reflect not only the changing makeup of Peru's destination markets but also changes in preferences within those markets. The fresh market tends to be more profitable than the processing market, but only in 2002 did fresh export volume finally rise above processed volume (fig. 4). In 2009, the fresh market accounted for 65 percent of export volume and the primary destination was the U.S. market, with 72 percent of all fresh asparagus exports. White asparagus is more popular in the EU and China, while green asparagus is the preferred choice in the United States.

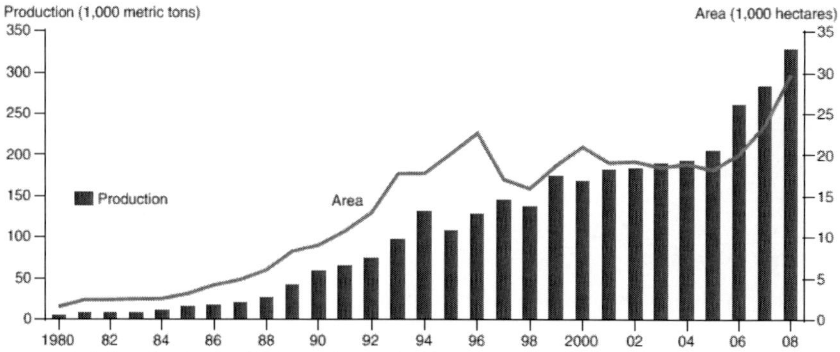

Source: Ministerio de Agricultura de Perú (2010).

Figure 3. Peruvian asparagus production and area, 1980-2008.

Processed and frozen asparagus accounted for 29 percent and 5 percent of export volume, respectively. Peru also exports processed and frozen asparagus throughout the year. In 2009, Peru exported 71 percent of their *processed* exports to the EU versus 23 percent to the United States (GTIS, 2010). Spain and France were the primary EU destinations. For the much smaller *frozen* asparagus industry, 53 percent of exports went to the United States and 40 percent to the EU in 2009.

Industry Structure

The Peruvian asparagus industry is dominated by large firms, but there are enough of them so that most claim only a small portion of total export share. In 2005, there were 122 asparagus exporters in Peru, 21/2 times as many as there were in 1995. The nation's five largest firms controlled approximately 42 percent of the market in 2005—the largest 50 firms shared 95 percent of the country's total fresh green asparagus output (Rios, 2007).

The Peruvian Institute for Asparagus and Horticultural Crops (IPEH) is a nonprofit association of asparagus producers and exporters founded in 1998 to promote research, marketing, extension, and international standard compliance. Grower organizations like IPEH have been critical to industry success since there is very little public agricultural research on fruit and vegetable production issues.

These organizations are an example of the entrepreneurship that is evident in the Peruvian agricultural export industry.

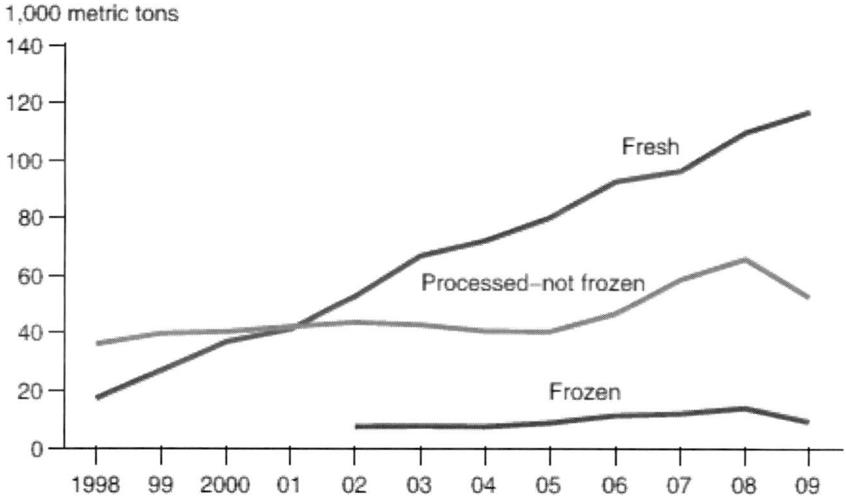

Source: Global Trade Information Services, Inc. (2010).

[1]In 2002, a separate Harmonized System code was first introduced for frozen asparagus. Before that, some frozen asparagus was exported, but it was reported jointly with other frozen vegetable exports.

Figure 4. Peruvian asparagus exports: Fresh, processed, and frozen, 1998-2009[1].

Tariffs and SPS Issues

While Peru's asparagus exports face no tariffs in their main markets, including the United States, sanitary and phytosanitary (SPS) barriers affect trade.

The U.S. Department of Agriculture's (USDA) Animal and Plant Health Inspection Service (APHIS) is trying to prevent entry of the moth *Copitarsia decolora*, which has been found on asparagus in Peru but not on asparagus in the United States.

APHIS requires every shipment of fresh green asparagus to the United States to be fumigated with the pesticide methyl bromide.[10] Fumigation is expensive and reduces the shelf life of asparagus, causing concern among Peruvian producers.

Potential Problems

Peru's growers recognize the inherent danger of relying on a small number of important export items. As the world's largest asparagus exporter, Peru runs the risk of oversupplying the market, which is one reason that most growers want to diversify crops. Transportation is a major cost component, and bottlenecks and rising transportation costs have become increasingly important issues. Most asparagus is exported by air freight. When air freight costs increase, some asparagus is shipped by boat, which is cheaper, but the lengthy transit reduces shelf life—a problem that becomes more acute with the recent addition of methyl bromide fumigation. Transportation costs and shelf life considerations will continue to determine whether long-distance shipments are economically viable.

Peru's main rival in the processed asparagus export business is China. Most Chinese exports consist of processed white asparagus that is shipped into the European market. China's main advantage comes from labor costs. Average production cost per hectare of asparagus in China is half that of Peru. Even taking into account Peru's higher yields, China still undercuts Peru on cost per ton—$317 versus $348 (TechnoServe Peru, 2005). While Peruvian exports enter the EU duty free and China is assessed a 16-percent duty, Chinese asparagus is still cheaper (GTIS, 2010). Dominance in the processed industry can change very quickly since no provider has a seasonal advantage and only the costs of production matter.

Impact on U.S. Production, Consumption, and Imports

Peru's success as a low-cost asparagus exporter played a large part in the recent decline of the U.S. asparagus industry. U.S. average annual asparagus production declined 58 percent between 1990-92 and 2007-09 (USDA, Economic Research Service (ERS), 2010a) (fig. 5). Different sectors of the asparagus industry are facing different levels of competition. In 1990-92, 59 percent of total U.S. asparagus production went to the fresh market, 32 percent to the canning market, and 9 percent to the freezing market. By 2007- 09, U.S. asparagus production had declined dramatically; fresh production was down 44 percent, and canned and frozen production had declined 84 and 57 percent, respectively. Increasing U.S. imports of fresh asparagus from Mexico in the spring have hurt the U.S. fresh industry, but Peru could play the same role if Mexican asparagus were not available. Peruvian competition was behind the decline in the canned and frozen asparagus industry (see box, "Peruvian Asparagus Exports: Impacts on California, Washington, and Michigan").

Peruvian Asparagus Exports: Impacts on California, Washington, and Michigan

The Fresh Market and California

California specializes in the fresh market and ships asparagus in the spring (box fig. 1). Although Peru is the largest source of U.S. fresh asparagus imports, it is spring imports from Mexico that have lowered prices and reduced early-season profitability for California growers. Competition with Mexico has led to a severe decline in the California fresh asparagus market; however, many growers confirm that, if Mexico were not supplying the market during early spring, Peru could step in with the same result. California asparagus production peaked in 2004 and, by 2008, had declined by 53 percent (box fig. 2). In 1990, the California season ran from late January to early May. In 2009, the season ran from March through early May.

Currently, Peruvian asparagus dominates in the second half of the year. Often, imports in the off season can raise domestic consumption as consumers become more familiar with a product and it becomes a menu staple, not just a seasonal treat. Between 1990-92 and 2007-09, average annual per capita U.S. consumption of fresh asparagus doubled. In the case of fresh asparagus, growers of U.S. imports, not U.S. producers, are benefiting from the increased consumption trends.

The Canning Industry and Washington

Competition from Peru has been most apparent in the canned asparagus market, which was largely concentrated in Washington. Imports first exceeded U.S. production in 2006 and by 2008 Peru accounted for 96 percent of U.S. processed asparagus imports. China was relegated from the most important import supplier to a distant second place with just 3 percent (GTIS, 2010).

The Washington asparagus season follows the California season, with shipments from early April through June. Traditionally, Washington growers could not rely on the early-season price premiums received by the California industry. They instead depended on high yields and concentrated on canned spears, which are particularly labor intensive since they require hand trimming. This reliance made their industry even more vulnerable to lower priced Peruvian imports. Washington asparagus production peaked at 51,200 tons in 1989 and by 2008 production declined 73 percent.

Between 2003 and 2005, the three biggest canners in the State (all multinational corporations) closed their operations and opened new plants in Peru. The now smaller Washington industry has been forced into the fresh market. As a fresh-market producer, Washington is facing less pressure. During its season, imports from both Mexico and Peru are low but could potentially increase. With lower production in the last few years during the Washington season, fresh-market asparagus prices have risen.

Frozen Asparagus and Michigan

Production patterns in Michigan have changed less than those in Washington, but Michigan growers have also been challenged by competition from Peru. Michigan traditionally focused on canned and frozen asparagus, particularly the product known as cuts and tips, which is less labor intensive than whole spears. Michigan asparagus production peaked in 2003 and, by 2008, had decreased by 19 percent.

Acreage reduction in Michigan is the result of the same factors as in Washington: import pressure with prices just above the cost of production. Michigan was able to avoid Washington's fate due to two factors. First, Michigan's production is more diversified, although canning is emphasized—approximately 45 percent of their production is canned, 35 percent is frozen, and 20 percent is sold as fresh. U.S. frozen asparagus production has been on a downward trend for decades, and imports first exceeded production in 2002. Peru has been the leading source of U.S. frozen asparagus imports since 1995 and supplied 82 percent of imports in 2008. China was the number two source, with 13 percent of the market in 2008 (GTIS, 2010).

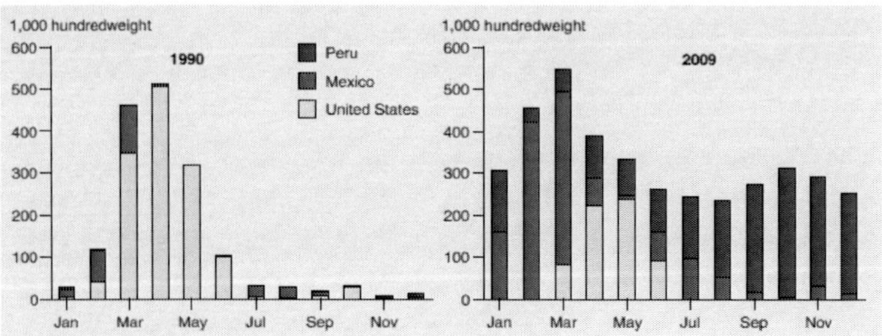

Source: USDA, Agricultural Marketing Service (2010).

Box figure 1. U.S., Mexican, and Peruvian fresh asparagus shipments, 1990 and 2009.

Michigan has the only remaining U.S. frozen asparagus industry. The second factor that helped Michigan maintain its processing industry is that many of the processing firms in Michigan are relatively small and locally owned, making them less likely to move operations to Peru than the large, multinational processors in Washington.

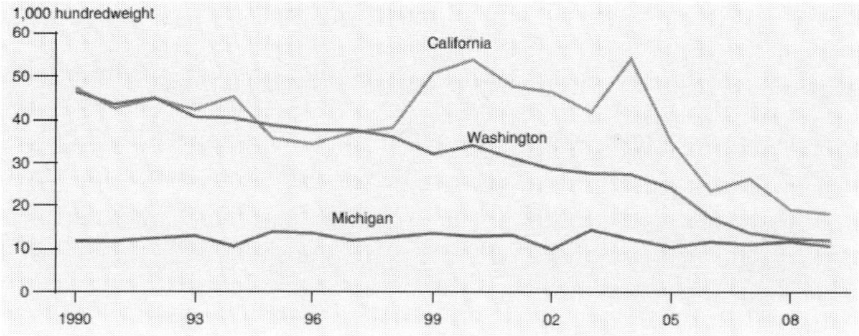

Source: USDA, National Agricultural Statistics Service (2010).

Box figure 2. Total U.S. asparagus production by State, 1990-2009.

Many are primarily private-label canners or freezers uninterested in outsourcing their operation because asparagus fills an important seasonal production niche. The Michigan Asparagus Advisory Board is encouraging producers to focus their efforts on production of fresh asparagus.

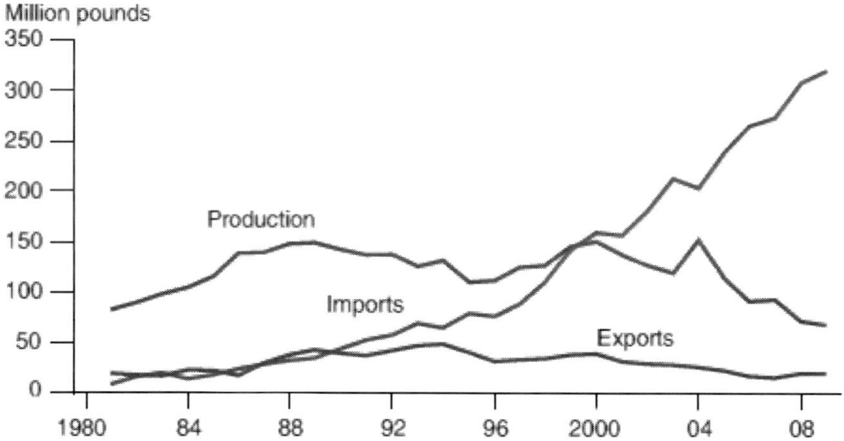

Source: USDA, Economic Research Service, 2010b.

Figure 5. U.S. fresh-market asparagus production, imports, and exports, 1980-2009.

In addition to competition from imports, changes in U.S. consumer preferences have affected the U.S. asparagus industry. During the last 20 years, U.S. average per capita consumption of canned asparagus fell more than 40 percent and frozen asparagus consumption remained unchanged. Aided by the new phenomenon of year-round supplies of fresh asparagus, average per capita consumption of fresh asparagus doubled.

The U.S. asparagus industry had long been protected from imports by high tariffs relative to those of other fruits and vegetables. Since implementation of the North American Free Trade Agreement (NAFTA) in 1994, U.S. imports from Mexico have increased as the tariff gradually declined to zero. Since 1991, Peruvian asparagus has entered the United States tariff free. As U.S. production declined, total U.S. fresh asparagus imports increased over 500 percent (fig. 5). Fresh imports first exceeded domestic production in 2000, and Peru has been the dominant supplier since 2002 when it surpassed imports from Mexico. Between 1990-92 and 2007-09, average annual imports of fresh asparagus from Peru increased from 2,835 metric tons to 78,644 metric tons. In 2009, Peru provided 56 percent of U.S. fresh imports, while Mexico supplied 43 percent. Frozen and canned asparagus imports also increased rapidly.

Federal Relief

U.S. producers will get some relief from the Asparagus Market Loss Program, part of the Food, Conservation, and Energy Act of 2008. A total of $15 million will be distributed to producers of fresh, processed, and frozen asparagus based on losses incurred from 2004 to 2007 due to increased imports. Although payments will help compensate growers for recent losses, many are concerned about whether domestic asparagus production can survive when the larger problem, depressed prices caused by increased foreign competition, remains. The Peruvian and Mexican asparagus industries have established themselves as forces on the world asparagus market.

Success Story #2: Processed Artichokes

While the success of the Peruvian asparagus industry had severe consequences for U.S. growers, the emergence of Peru's processed artichoke industry had no effect on U.S. producers. U.S. growers were not hurt because the United States produces only fresh artichokes; U.S. growers abandoned the processing market because it was not competitive with Spain's exports. When

Peru entered the market, its growing exports negatively affected other suppliers to the United States, most notably Spain.

Production

Artichokes were grown in Peru before the boom in fruit and vegetable exports but only on a very small scale. Large-scale production for export began after disease seriously affected the country's asparagus crop between 1998 and 1999, and growers sought to diversify their exports (Vallejos, 2004). On the coast, growing conditions allow for a 4- to 5-month harvest season. Unlike most of Peru's other exported crops, artichoke production is not restricted to the country's arid coastal desert. Artichokes thrive in some areas of the Sierra. Several of the large artichoke producers on the coast tried to expand production to the Sierra, but they encountered many problems, such as limited grower technical skills, more variable weather, and insufficient transportation infrastructure. Several development organizations are currently promoting small-scale production in the Sierra in order to provide employment and income opportunities to impoverished communities.

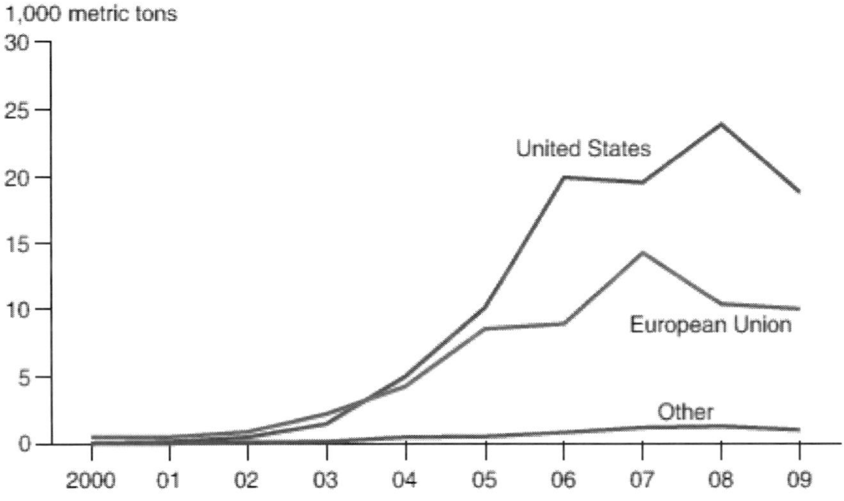

Source: Global Trade Information Services, Inc. (2010).

Figure 6. Peruvian processed artichoke exports by destination, 2000-2009.

Exports

After a 1999-2001 trial period, processed artichoke exports increased rapidly, reaching 35,000 metric tons in 2008 (fig. 6). In 2009, artichokes were

Peru's fifth highest value fruit and vegetable export commodity at $72.3 million (GTIS, 2010). Almost all of Peru's artichoke exports are processed products. The large export firms of the Peruvian coast concentrate their production on varieties that produce the best processed product, primarily artichoke hearts. These processing varieties are not the familiar green globe variety grown in the United States for the fresh market. Peru does, however, occasionally export a very small amount of fresh artichokes to the United States in the fall if U.S. prices are sufficiently high (the last shipment was in 2007).

The United States, taking 60 percent of Peru's artichoke export, is Peru's main market, followed by Spain and France, each taking 21 and 8 percent, respectively (GTIS, 2010). Processed artichokes are Peru's second largest fruit and vegetable export in value to the United States. Peru's main competitors are Spain and Chile, although Peru has been gaining part of Spain's market share in the past few years. Artichoke exports to the United States benefit from completely liberalized trade and the total lack of SPS issues. Fresh globe artichokes are permitted entry into all U.S. ports without any special treatment processes, and all processed artichokes are permitted entry.

Impact on U.S. Production and Imports

Almost 99 percent of artichoke acreage planted in the United States today is in California. U.S. artichoke production has vacillated around 100 million pounds for the past decade after 1985's production peak of 134 million pounds. In the past, California growers sent artichokes to the fresh and processing market, but the one processor went out of business in 2000, partly because of competition from Spain. Now, all artichokes go to the fresh market. U.S. fresh artichoke producers are seeing little competition from imports, which account for just 2 percent of fresh consumption, mostly from Mexico.

Since the U.S. artichoke-processing industry was gone before Peru became a major player, Peru has taken market share away from other foreign processors. Imports of processed artichokes from all sources have increased 242 percent since 1990. U.S. processed artichoke imports from Peru increased rapidly from the first shipments in 2001 (fig. 7). While total processed artichoke imports increased 41 percent since Peru began exporting, much of Peru's supply has substituted for U.S. imports of Spanish processed artichokes, which have declined precipitously. In 2009, Peru accounted for 45 percent of U.S. processed artichoke imports, followed by Spain with 26 percent, and Chile with 19 percent (GTIS, 2010).

If artichoke-processing countries, such as Peru and Chile, improve quality standards and successfully transition into fresh artichoke production, then U.S. producers could be affected by the threat of cheaper imports. For now, however, processed Peruvian artichokes have benefited the U.S. market by increasing available supplies and decreasing prices for consumers.

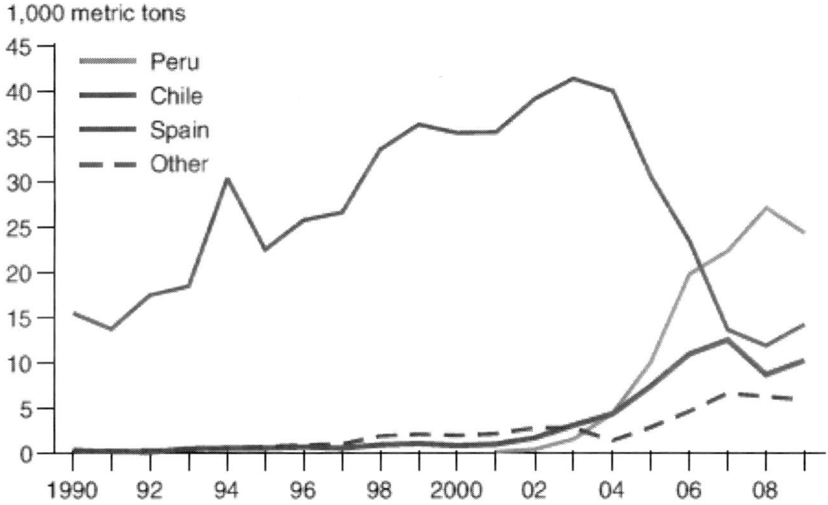

Source: Global Trade Information Services, Inc. (2010).

Figure 7. U.S. processed artichoke imports by country of origin, 1990-2009.

Success Story #3: Table Grapes

Whereas asparagus represents products for which the United States and Peru compete in the U.S. market and processed artichokes do not, U.S. and Peruvian fruit and vegetable products can compete in another area: third-country export markets. Table grapes are a good example of third-country competition between the United States and Peru. Although Peru's season is mostly opposite that of the United States, partly overlapping production seasons have caused competition in exports to other countries to rise.

Production

In the late 1990s, Peruvian agribusiness began to investigate the potential of table grapes as a profitable export. Until that time, grapes had been grown on a smaller scale, destined principally for the domestic table grape market, as well as for wine and spirits production. The Peruvian table grape export industry is still relatively young, experimenting with different production areas and varieties. Land is still available that could be brought into grape production but that might not have adequate soils for other crops (MINAG, 2008). Thus, the full potential of the industry may take some time to develop. Grape production in Peru is concentrated in the coastal regions of Ica (42 percent), Lima (26 percent), and La Libertad (23 percent) (MINAG, 2010). New production areas are developing in Piura and Lambayeque, areas in the north that have a very early season, thanks to that region's warmer climate.

Production is currently concentrated on four varieties—Red Globe (with 75 percent of production), Flame Seedless (9 percent), Sugraone (8 percent), and Crimson Seedless (5 percent) (USDA, Foreign Agricultural Service (FAS), 2009). Peru began with the seeded Red Globe grape, which grows well there. It is a thick-skinned grape that is relatively easy to ship compared with other varieties.

However, much of the growth in grape demand worldwide is for seedless varieties, and Peru is responding to this demand trend. Not all varieties produce equally well in Peru; Thompson seedless, the dominant U.S. table grape, has not yet shown much success in Peru (Boriss, 2006).

Exports

Since the first 600 metric tons of table grapes Peru exported in 1998, the industry grew to over 60,000 metric tons exported in 2009 (fig. 8). Peru exported $127 million in grapes to all countries in 2009, making grapes the country's second most valuable fruit and vegetable export, with $31.6 million exported to the United States. Although Peru is not a major grape producer or exporter, only six of the world's top 25 exporters—Chile, South Africa, Argentina, Brazil, Australia, and Peru—are located in the Southern Hemisphere, meaning that Peru is one of only a handful of countries that can supply markets during the Northern Hemisphere's winter. Peru exports grapes mainly from October to March, with the greatest volume from November to February. This export window falls at the end of the U.S. grape season and the beginning of the Chilean season. In addition, the height of Peru's production season coincides with the strong demand associated with worldwide December holiday celebrations, as well as the Chinese New Year.

With 32 percent of volume in 2009, Europe is Peru's biggest export market, followed by the United States with 23 percent, and Northeast Asia (comprised of Hong Kong, China, Taiwan, Japan, and South Korea) with 19 percent. Grapes can be stored and are exported by ship.

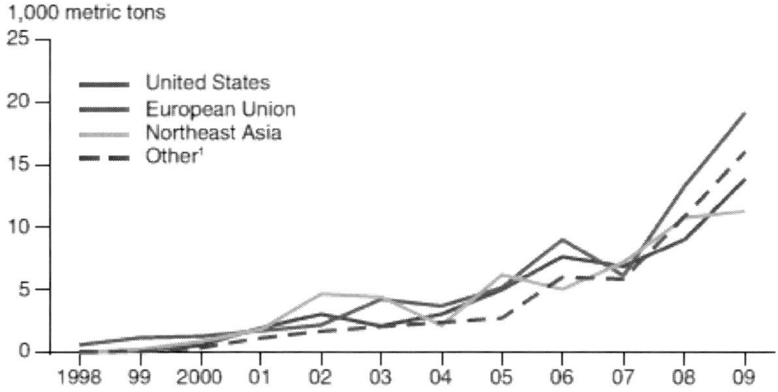

Source: Global Trade Information Services, Inc. (2010).

[1] No country or aggregation of countries is bigger than the EU, United States, or Northeast Asia.

Figure 8. Peruvian grape exports by destination, 1998-2009.

Tariffs and SPS Issues

Peruvian grapes enter the U.S. duty free, as do grapes from the two other major U.S. suppliers, Chile and Mexico. On the phytosanitary front, table grape imports from Peru are permitted but must be cold treated in order to prevent the entry of the Mediterranean and various other fruit flies into the United States (USDA, APHIS, 2009). Imports are also restricted to certain ports.

Industry Structure

The table grape export industry is highly concentrated and dominated by the 36 member firms of PROVID, the Peruvian grape producer association. Three Peruvian companies account for approximately 50 percent of table grape exports (USDA, FAS, 2009). The entire membership of the organization now accounts for around 90 percent of Peru's table grape exports (PROVID, 2008). Founded in 2001, the organization is working to open new markets for its members, especially the lucrative Northeast Asian countries of Japan and South Korea. In this pursuit, the association has helped its members adopt stricter phytosanitary standards that benefit the country's industry as a whole and better position Peru's producers for global competition.

Impact on U.S. Production, Imports, and Exports

California is the source of virtually all U.S. table grape production. Annual U.S. table grape production has averaged nearly 1 million tons since the late 1990s (fig. 9). Both imports and exports have grown. Traditionally, the U.S. season ran from May through December; new late-season varieties, such as Red Globe, Crimson Seedless, and Autumn Royal, now enable shipping into January.

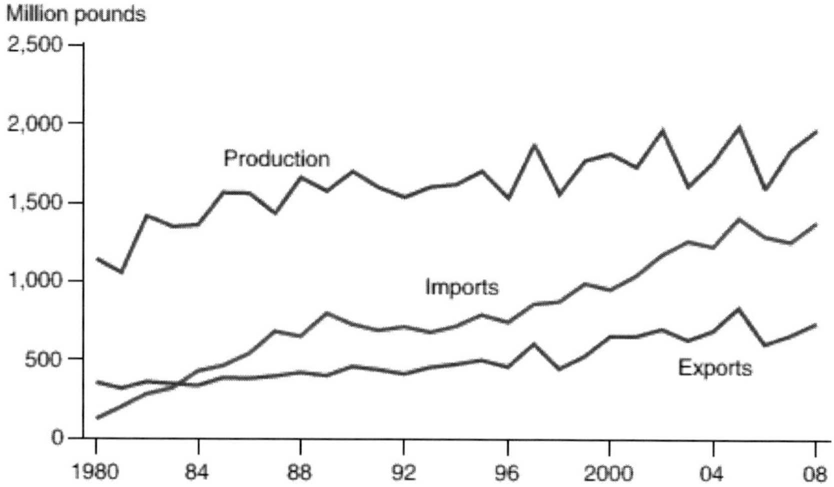

Source: USDA, Economic Research Service (2010c).

Figure 9. U.S. table grape production, imports, and exports, 1970-2008.

The majority of U.S. off-season grape imports are from Chile and Mexico, with small amounts from other suppliers, including Peru, from November through March (fig. 10). Peru is currently not positioned to increase market share substantially in the United States unless it develops a market niche by offering a type or quality of grape that is not available in a particular period or by offering grapes at a lower price than that of the existing competition. Peru's share of the U.S. market totaled just 3 percent in 2009; as U.S. growers find ways to extend their season, Peru's market niche in the United States may decline even further.

Peru may have a better chance to take market share in the United States from its regional competitors. Brazil benefits from a short market niche between September and November. For several years, very few U.S. green grapes were available in October and November, and Brazil could demand a premium for its exports. If Peru can produce the type of green grape that Brazil exports to the United States at a lower cost than Brazil does, Peru may be able to compete successfully. U.S. consumers would then benefit from lower prices. Peru overtook Brazil in 2009 as the third largest grape exporter to the United States after Chile and Mexico. Peru might also pose a challenge to Chile's very early season exports, which have high production costs.

Exports are very important to the health of the U.S. table grape industry. In the 2008/09 marketing season, U.S. grape exports represented 38 percent of production. In 2009, the largest export markets were Canada, Mexico, and China/Hong Kong, which accounted for 45 percent, 8 percent, and 10 percent, respectively (table 2). The United States exported a relatively small amount of grapes to 78 other countries; even a small increase in grapes from Peru could affect these markets.

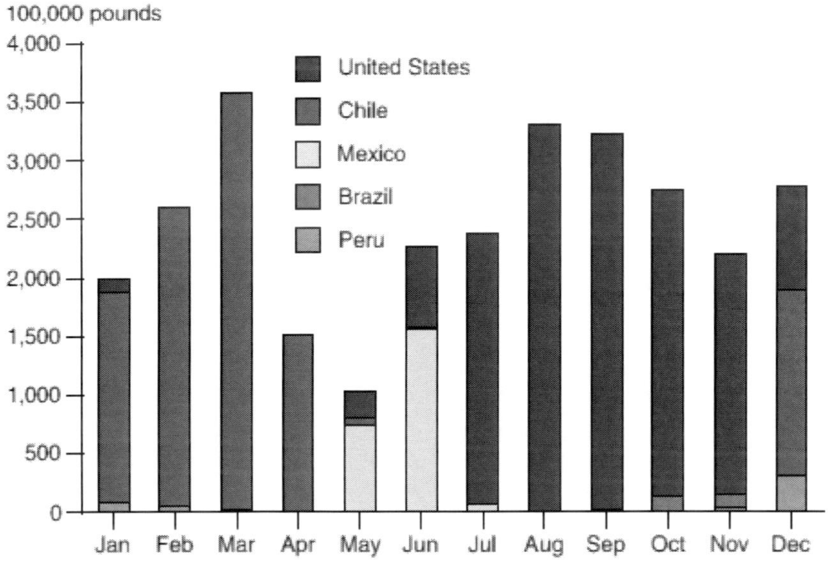

Source: Agricultural Marketing Service, 2010.

Figure 10. U.S. table grape shipments by source, 2009.

Peru has the potential to be a competitor of concern in third countries. For example, China/Hong Kong is an important market for the United States and Peru, and both countries are trying to capture the lucrative Chinese New Year market. U.S. exports have been declining in the last decade, whereas Peru's exports, although small, have been growing (fig. 11). The United States is the largest source of imports in China/Hong Kong from August through January, but Peruvian imports are now competing, particularly in January (fig. 12). Peru is also supplying a large share of the grapes in February. The United Kingdom is another important market for both the United States and Peru (3 percent and 6 percent of their exports, respectively) (table 2). Peruvian exports to that market have been growing steadily. As Peru continues to improve the quality of their grapes, consumer preferences for various grape varieties in third-country markets will likely determine relative market dominance between the United States and Peru.

Table 2. U.S. and Peruvian table grape exports to selected third-country markets, 20091

Ranking	Country or region	United States		Peru	
		Quantity	Share	Quantity	Share
		Metric tons	*Percent*	*Metric tons*	*Percent*
1	Canada	167,424	45	591	1
2	Hong Kong	30,483	8	7,299	12
3	Mexico	16,392	4	0	0
4	Indonesia	16,129	4	1,575	3
5	Australia	15,397	4	19	0
6	Philippines	14,552	4	187	0
7	Taiwan	14,297	4	856	1
8	United Kingdom	9,775	3	3,610	6
9	Malaysia	7,823	2	333	1
10	Thailand	7,115	2	899	1
11	New Zealand	6,167	2	19	0
12	Singapore	6,092	2	188	0
13	China	6,083	2	3,112	5
14	Guatemala	5,836	2	168	0
15	Vietnam	5,534	1	991	2
16	Colombia	2,937	1	2,063	3
17	Russia	2,322	1	5,244	9
18	Netherlands	494	0	10,902	18
19	Spain	0	0	2,665	4
	Others[2]	40,468	11	19,413	32

Source: Global Trade Information Service, Inc. (2010).

[1]Although this table lists annual market shares, please note that the United States competes with Peru mostly during December and January.

[2]The largest share of Peru's fresh grape "Others" exports goes to the United States at 23 percent.

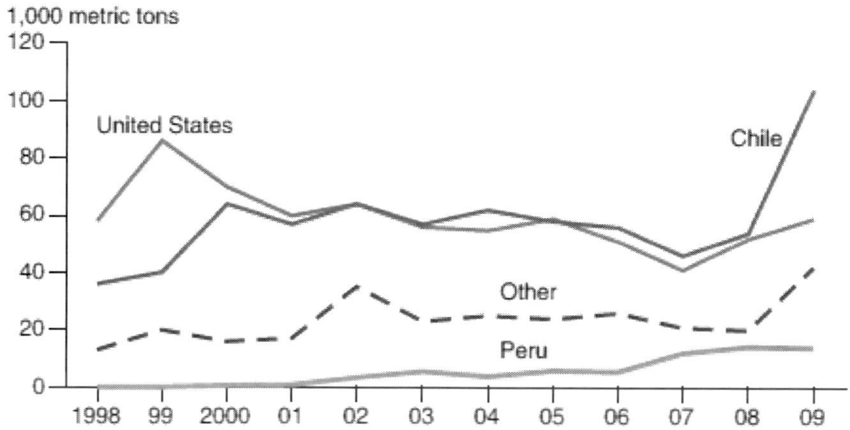

Source: Global Trade Information Services, Inc. (2010).

[1]Although this figure lists annual market shares, please note that the United States competes with Peru mostly during December and January.

Figure 11. China/Hong Kong table grape imports by country of origin, 1998-2009.[1]

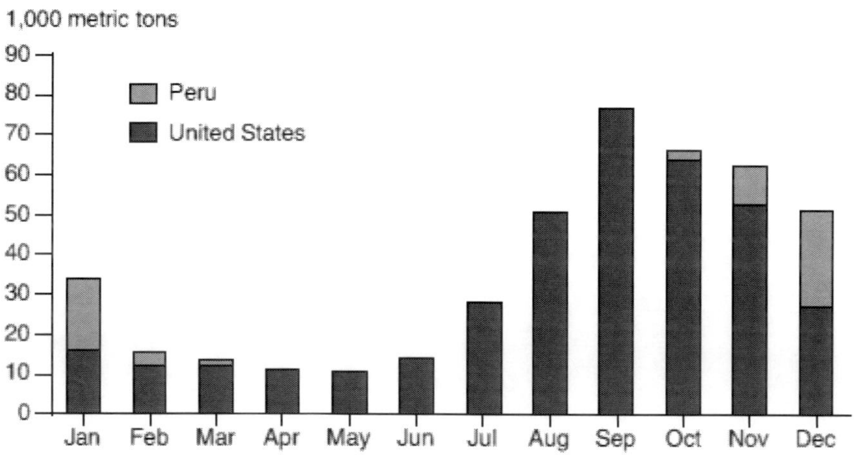

Source: Global Trade Information Services, Inc. (2010).

Figure 12. Overlap in monthly China/Hong Kong grape imports from Peru and the United States, 2009.

Challenges for Peru's Fruit and Vegetable Export Industry

Peru's continued success as a fruit and vegetable exporter faces a number of challenges. Reliance on a small number of export products increases the sector's vulnerability to international market shocks. Also, the great distance between Peru and its major trading partners makes transportation costs a crucial factor in competitiveness, and export infrastructure, such as roads and port capacity, is not keeping pace with rapidly growing export volumes. Finally, old problems such as water management and land rights also continue to burden the fruit and vegetable export industry.

Reliance on a Small Number of Products and Markets

Asparagus still dominates Peru's horticultural export success. A country that depends on a narrow export portfolio, however, is more vulnerable to shocks. There is no or very limited domestic market demand for some of these products, which puts even greater importance on export markets. Growers, therefore, try to diversify into other crops by experimenting with new export commodities. Once new promising export products are found, however, a lengthy process may be necessary to get them cleared for export to the United States or other countries. APHIS, concerned about the introduction of non-native pests, conducts extensive investigations before it opens the U.S. market to new foreign products, particularly fresh products. Peru is still lagging behind its neighbor Chile, a long-time fruit and vegetable exporter, which has APHIS approval for many more horticultural products than Peru.

Besides efforts to diversify its range of export products, Peru is also trying to diversify its export markets. This effort is particularly important for crops like asparagus for which inadvertent overproduction can lower prices to unprofitable levels. The increasing number of FTAs is a testimony to this effort of market diversification.

Distance

Peru is a long way from the United States and other important markets, and transportation is therefore a sizable cost factor. As a result, Peru must concentrate on high-value commodities. Asparagus was traditionally sent to the U.S. market by air freight. In the early days of the industry, asparagus prices were very high. As export volume increased and prices fell, the cost of air freight became a challenge to the industry. In recent years, air freight accounted for about 45 percent of its total delivered cost. Ocean freight is considerably cheaper than air freight, but more time in transit reduces the product's shelf life. Both air freight and ocean freight costs have increased substantially since the early 1990s, peaking in 2008, when high energy costs raised air freight for U.S.-bound exports by 15 percent over 2007 levels. High transportation costs make Peru vulnerable to competition from suppliers closer to the United States, such as Mexico.

Export Infrastructure

The lack of adequate infrastructure is one of the most serious constraints to Peruvian export expansion. Peru requires substantial investments to improve its roads and ports, and the Government of Peru is working toward this goal. The Pan-American Highway traverses the country from north to south along the coast. Although portions of this road around Lima have four lanes, the vast majority of the highway in Peru is just two lanes, which results in congestion during harvest seasons. Most of the east-west roads to the Sierra are even more inadequate—a serious impediment for growers who might want to expand production to areas beyond the coast.

When the asparagus export industry emerged, growers struggled with inadequate airport infrastructure to maintain the cold chain, which is critical for a perishable product. Growers banded together in 1998 to form Frio Aéreo Asociación Civil, a private nonprofit association, to improve the logistics for fresh asparagus exports by air. This association has been very effective and offers services like cold storage, quality control, and fumigation. As a result, the air freight export process for asparagus and other produce is now much smoother.

Some commodity exports, including fresh grapes and processed vegetables, generally ship by sea. Rapidly growing export volumes of these products have placed pressure on Peruvian port facilities. Callao, just south of Lima, is by far Peru's most important port, handling close to 50 percent of Peru's entire trade volume in 2007, but it is also one of the most expensive ports in the world. The average cost of exporting one container from Peru is $800, compared with an average of $510 in Chile and $660 in Costa Rica. Aside from port capacity, insufficient port logistics (such as insurance, holding, or security) also put Peru at a disadvantage. In 2004, Peru's logistics costs were 32 percent of product value, compared with 9.5 percent in the United States and 18 percent in Chile (Gonzalez et al., 2007). The Peruvian Government encourages private and foreign investment in infrastructure projects, such as the expanding and improving of port facilities, but some workers' unions oppose foreign investment (*Reuters*, 2009b).

Water Management

Peru has ample supplies of renewable fresh water—more per capita than any other country in Latin America—but water distribution is a problem. The Sierra and the jungle receive abundant rains, but the Pacific watershed, where the primary fruit and vegetable exporters operate (and 50 percent of the population live), accounts for only 2 percent of available surface water (Amat y León, 2006).

In the coastal region, water is available from about 60 rivers that flow from the Sierra to the Pacific, although the flow varies seasonally. Agriculture is concentrated in these river valleys. New irrigation projects, using river water, have opened some desert areas on the coast to agriculture. In addition, some fruit and vegetable exporters based in the coastal desert area depend on groundwater.

Since the 1950s, Government investment has brought approximately 1 million hectares on the coast under irrigation, using mostly surface water, 80 percent of which is used to irrigate agricultural production. However, expanded agricultural area, poor infrastructure, and inefficient irrigation methods have led to dwindling water supplies. The more technologically advanced exporters have introduced more water-saving irrigation technologies, such as drip irrigation, but at the same time, the amount of irrigated land is increasing and multiple cropping cycles per year are becoming more common, raising irrigation water demand.

Some of the most important areas for fruit and vegetable export production are suffering from overexploitation of groundwater and salinization. The extraction of groundwater in some parts of the country, such as the Ica Valley, 200 kilometers (km) south of Lima, is progressing rapidly (Olson, 2006). Water supplies in Ica are projected to be exhausted possibly within 15 years unless dramatic action is taken (Rendón Schneir, 2009).

In recent years, the Government of Peru has begun to reform water regulations. At the same time, the Government has also created several new institutions that have a role in water resources management.[11]

Incomplete water rights, as well as the lack of appropriate pricing for water use, are obstacles to investment in irrigation projects. Currently, producers pay just a fraction of the true cost of water—$30 per hectare per year for a medium-sized farm in a coastal valley, when the true cost can be $200 (USDA, FAS, 2007). The Government of Peru plans to increase the area under drip irrigation to 200,000 hectares on the coast (up from 70,000 in 2007) and establish 30,000 hectares with drip irrigation in the Sierra. The World Bank is funding a project that aims to aid the Government in its design of use and discharge fees, and it plans to help the National Water Authority (ANA) set up local river basin offices.

Even if existing water resources are managed efficiently, they still may not be sufficient to supply growing needs in the coastal area. Peruvians talk about tapping the more-than-sufficient water supplies on the eastern side of the Andes and bringing them to the Pacific watershed. The Government is encouraging public-private partnerships to raise the considerable funds necessary for this kind of investment. One example of a successful public-private investment partnership is the Chavimochic irrigation project in La Libertad, 500 km north of Lima, one of Peru's major agricultural production areas. Water provided by the Chavimochic project is sold at a price to defray the cost of the irrigation system, which consists of pipes that run through mountains in order to tap water from a larger river to the south. This redirected water is then distributed through a system of channels.

Land Rights Issues in Peru

Land rights are another salient issue affecting Peru's agriculture export sector. Land reform was implemented in Peru between 1969 and 1975 as a means to achieve greater social justice and to alleviate rural poverty. Before these reforms, the 1961 Peruvian census showed that 75.6 percent of the surveyed land was owned by 0.4 percent of the population (Eguren, 2006). After the Third Agrarian Reform Law of 1969, expropriation and redistribution of land did not bring the intended benefits. The land originally given to agricultural production cooperatives was subdivided and sold to their previous members. While these subdivisions did give land directly to smallholders, the transfers were problematic as most were not accompanied by a legal government-issued title. The lack of well-defined property rights was underscored by the 1994 census. Of the country's 5.7 million farms, 53 percent were not titled (Herrera, 2005).

With the passing years and administrations, the discourse over land rights in Peru has changed from a focus on how "the land belongs to the person who works it" toward an emphasis on efficiency and competitiveness, which often requires a sufficiently large farm size and experienced and active management. President Alberto Fujimori solidified decrees to this effect with the enactment of a new constitution in 1993 that allowed for transfer of land titles to third parties and eliminated all acreage restrictions on landholdings. The Titling Act from 1997 went even further as it contained a provision that uncultivated land holdings would be declared "abandoned" and taken over by the State. This new law created some tension with small landholders who left some of their coastal lands fallow because of lack of water resources and not as the result of land abandonment. Many of the largest farms in Peru acquired their land from the Government after just such a takeover (La Revista Agraria, 2009).

The result of this new land distribution process has been increased concentration of acreage by just a handful of firms producing crops predominantly bound for export. The expansion of arable lands into the desert—made possible by large investments in irrigation schemes financed so far with mostly public funds—is likewise dominated by export firms. More than three-quarters of the 66,000 hectares of land opened up by the Chavimochic irrigation project were bought by 11 investors.

Challenges remain despite the recent success of agricultural production destined for exports. Many larger farms are limited in their potential to expand. A large share of small-scale growers still lack titles for their land and are thus prevented from selling it even though progress has been made in the area of providing land titles.

CONCLUSION

The fruit and vegetable export industry in Peru has expanded rapidly over the last 20 years and has made Peru an important player in world markets for a number of commodities. Peru is the world's leading asparagus exporter, and other horticultural products may follow suit. Several factors contribute to Peru's success: a favorable business environment, trade agreements, low labor costs, and a climate that favors production of many fruit and vegetable products. Peru was able to capture significant market share in both the United

States and Europe (and occasionally Northeast Asia) for a variety of fruit and vegetable products. This increased market share has had varied impacts on U.S. production and exports. In some instances, it has displaced domestic industries, as was the case in the processed asparagus industry. In other cases, such as processed artichokes, new Peruvian supplies in the U.S. market do not compete with U.S. growers. In still other instances, such as in the table grape trade, the United States finds itself competing with Peru in certain months for third-country markets, such as Northeast Asia and the United Kingdom. Peru's continued success in the sector will require overcoming challenges, such as infrastructure bottlenecks, risk exposure due to narrow export portfolios, water management issues, and a complicated land tenure situation.

REFERENCES

Amat y León, Carlos (2006). *El Perú nuestro de cada día*, Universidad del Pacifico, Lima, Peru.

Andina (Agencia Peruana de Noticias) (2010). "Peru's free trade pact with China to boost exports," February. 28, www.andina.com. pe/ingles/ Noticia. aspx?id=49fInhVg+s8=.

Asparagus Growing (2010). "Harvest asparagus," www.asparagusgrowing. net/asparagus-growing/harvest-asparagus.

Benson, B.L. (2010). *Update of the World's Asparagus Production Areas, Spear Utilization, and Production Periods*, California Asparagus Seed and Transplants, Inc., Davis, CA.

Bolle, Mary Jane, and M. Angeles Villarreal (2007). *Peru Trade Promotion Agreement: Labor Issues*, Order Code RS22521, Congressional Research Service, Foreign Affairs, Defense, and Trade Division, updated December 18, 2007.

Boriss, H. (2006). *Commodity Profile: Table Grapes*, Agricultural Marketing Resource Center, University of California, Davis.

The Economist Intelligence Unit (1997 and 2008). Country Profile Peru, http://www.eiu.com/index.asp?layout=displayIssue&publica tion_id=980000698.

The Economist Intelligence Unit (2010). Country Report Peru, February, http://www.eiu.com/index.asp?layout=displayIssueTOC&toc2=no&i ssue_id=955231680&publication_id=880000888.

Eguren, Fernando (2006). *La Reforma Agraria en el Perú*, Food and Agriculture Organization of the United Nations, Centro Peruano de Estudios Sociales, Rome, December.

Flores, Carlos Necochea (2007). "La agricultura andina esta en riesgo por el acelerada deshielo de las montañas," El Comercio.com.pe, October 18.

Food and Agriculture Organization of the United Nations (2009). FAOSTAT database, Rome, faostat.fao.org.

Global Trade Information Services, Inc. (2010). Global Trade Atlas, http://www.gtis.com/gta/.

Gonzalez, Julio A., Jose Luis Guasch, and Tomas Serebrisky (2007). *Latin America: Addressing High Logistics Costs and Poor Infrastructure for Merchandise Transportation and Trade Facilitation*, World Bank, Washington, DC, August.

Herrera, M. (2005). "*El Estado de la Información sobre Tenencia de al Tierra para la Formulación de Políticas de Tierra in América Latina*," Food and Agriculture Organization of the United Nations, Rome, www.oas.org/dsd/spanish/Documentos/Informacionsobretenenciadelatierraypoliticasdetierra.pdf.

La Revista Agraria (2009). *Especial: Concentración de tierra*, p. 11, Centro Peruano de Estudios Sociales (Peruvian Center for Social Studies), Lima, Peru, May.

Ministerio de Agricultura del Perú (2010). Series Históricas de Producción Agrícola – Compendio Estadístico, frenteweb.minag.gob.pe/sisca.

Ministerio de Agricultura de Perú, Dirección General de Información Agraria (2008). *Informe de registro de productores de uva en las regions de Ica, Arequipa, Moquegua, Tacna y Lima provincias*, September, www.minag.gob.pe/download/pdf/herramientas/boletines/DocumentoFinalVid.pdf.

Nolte, Gaspar (2007). *Peru Asparagus Annual*, GAIN Report No. PE7011, U.S. Department of Agriculture, Foreign Agricultural Service, June 8, www.fas.usda.gov/gainfiles/200706/146291392.pdf.

Olson, Douglas (2006). "Chapter 18: Recursos hídricos," Marcelo M. Giugale, Vincente Fretes-Cibils, and John L. Newman (eds.), *Perú, La oportunidad de un país diferente: próspero, equitativo y gobernable*, World Bank, Washington, DC.

PROVID (Asociación de Productores de Uva del Perú) (2008). *PROVID: 7 años al servicio del agroexportador visionario*, Lima, Perú.

Rendón Schneir, Eric (2009). "Agroexportación, desempeño ambiental y propuesta de manejo sostenible de los recursos hídricos en el valle de Ica:

1950 – 2007," Draft for dissertation, Universidad Nacional Autónoma de México.

Reuters (2009a). "Update 2—Peru central bank cuts GDP view for 2009, 2010," September 18, http://www.reuters.com/assets/ print?aid= USN183 1821620090918.

Reuters (2009b). "Peru port workers start strike over privatization," November 5, www.reuters.com/article/marketsNews/ idUSN0541587620091105.

Rios, L. (2007). *Agro industries characterization and appraisal: Asparagus in Peru*, Food and Agriculture Organization of the United Nations, Rome.

TechnoServe Peru (2005)."Visión del Esparrago Peruano," tns.org.pe/peru/ visiones/Esparragos%202005.pdf.

U.S. Department of Agriculture, Agricultural Marketing Service (2010). *Fresh Fruit and Vegetable Shipments by Commodities, States, and Months*, FVAS-4 Calendar Year 2009, February, http://www.ams.usda.gov/ AMSv1.0/getfile?dDocName=STELPRDC5082945.

U.S. Department of Agriculture, Animal and Plant Health Inspection Service (2009). Fruits and Vegetable Import Requirements (FAVIR), Commodity Import Report (CIR): Grape (Fruit, or cluster of fruit) from Peru into All Port(s), https://epermits.aphis.usda.gov/manual/index. cfm?action= cir ReportP&PERMITTED_ID=8264.

U.S. Department of Agriculture, Economic Research Service (2010a). Table 6—Vegetables, fresh market: Production of major vegetables, 1980-2009, Vegetables and Melons Yearbook Dataset, http://usda.mannlib.cornell. edu/ usda/ers/89011/Table006.xls.

U.S. Department of Agriculture, Economic Research Service (2010b). Table 64—U.S. fresh asparagus: Supply, utilization, and price, farm weight, 1970-2010, Vegetables and Melons Yearbook Dataset, http://usda. mannlib. cornell.edu/usda/ers/89011/Table064.xls.

U.S. Department of Agriculture, Economic Research Service (2010c). Table B-14—Grapes: Production, utilization, and season-average grower price, United States, 1980 to date, Fruit and Tree Nut Yearbook Spreadsheet Files, usda.mannlib.cornell.edu/usda/ers/89022/Table-B14.xls.

U.S. Department of Agriculture, Foreign Agricultural Service (2007). *Peru Agricultural Situation Update 2007*, GAIN Report No. PE7018, August 15, www.fas.usda.gov/gainfiles/200708/146292068.pdf.

U.S. Department of Agriculture, Foreign Agricultural Service (2009). *Peru Grape Report*, Gain Report, August 28, gain.fas.usda.gov/Recent%20 GAIN%20Publications/Grape%20Report_Lima_Peru_8-28-2009.pdf.

U.S. Department of Agriculture, National Agricultural Statistics Service (2010). *Vegetables Annual Summary*, January, http://usda.mannlib.cornell.edu/usda/current/VegeSumm/VegeSumm-01-27-2010.pdf.

U.S. Trade Representative (2009). Peru Trade Promotion Agreement, Washington, DC, June, www.ustr.gov/trade-agreements/free-trade-agreements/peru-tpa.

Vallejos, Francisco (2004). "Exportaciones de alcachofas superan a espárragos en zona de La Libertad," *El Comercio*, Lima, Peru, November 11.

World Bank (2009a). *Peru Water Resources Management Modernization Project*, Report No. AB4447, Project Information Document (PID) Appraisal Stage, Washington, DC, January.

World Bank (2009b). *World Development Indicators 2009*, Washington, DC.

World Bank (2010). *World Development Indicators 2010,* Washington, DC.

End Notes

[1] Peru began to feel the effects of the worldwide economic downturn in 2009, with growth projected at only 1.8 percent, recovering to 5 percent in 2010 (Reuters, 2009a).

[2] Many mineral commodity prices sharply declined after 2007 until 2009 when they began to recover.

[3] Examples for such associations are the Instituto Peruano del Esparrago y Hortalizas, which promotes export interests for asparagus and artichoke producers, as well as ProHass, the association of Hass avocado producers, and ProCitrus, the association for Peruvian citrus producers.

[4] In 2005, 44 percent of all U.S. imports from Peru, mostly copper and certain apparel, received preferential duty treatment (Villarreal, 2006).

[5] Assuming 180 hours of work per month.

[6] For more information on asparagus harvesting, see Asparagus Growing (2010), http://www.asparagusgrowing. et/asparagus-growing/harvest-asparagus.

[7] These issues are discussed at length in the "Water Management" section.

[8] White asparagus is a more labor-intensive crop because the emerging spears must be covered with a mound of soil to keep the sun from turning them green.

[9] Yields vary by region. In La Libertad, average yield in 2007 was 13.4 tons/hectare (ha) (Ministerio de Agricultura del Perú, 2010). Under the right conditions and with the most efficient, sophisticated technologies, yields as high as 20 tons/ha have been reached (Nolte, 2007).

[10] White asparagus is not fumigated because it develops under the soil and is therefore not a host to the pest.

[11] The Government of Peru created the National Water Authority (Autoridad Nacional del Agua—ANA) as part of the Ministry of Agriculture. This agency enjoys financial and administrative autonomy and has a clear mandate for integrated and multisectoral water resource management. The Government also created a new Ministry of Environment,

which, among other tasks, is responsible for generating hydrological information. Furthermore, the Ministry of Health is involved in water issues as it is responsible for water quality management. However, having three ministries in charge of water issues has led to confusion and conflict over each institution's set of responsibilities. In May 2008, the Ministry of Environment was put in charge of protecting water resources overseeing their recuperation, and the Ministry of Agriculture was put in charge of distributing water permits. Another concern is the potential conflict of interest from ANA being part of the Ministry of Agriculture while agriculture is one of the major water users (World Bank, 2009a).

In: Global Agriculture. Volume 1 ISBN: 978-1-62257-369-1
Editors: M.R. Robertson, J.I. Schmidt

Chapter 2

THE U.S. AND MEXICAN DRY BEAN SECTORS*

Steven Zahniser, Marcela Vera Torres, José Alberto Cuéllar Álvarez, Nicolás Fernando López López, and Rhea Bhatta

ABSTRACT

Liberalization of U.S.-Mexico dry bean trade, as part of the North American Free Trade Agreement (NAFTA), has enabled U.S. dry bean producers to provide a steadier and larger portion of Mexican supply. Long-term prospects for the newly integrated market are shaped by various factors.

On the supply side, the future structure of Mexico's dry bean sector will be comprised of a smaller number of producers who work larger plots of land for higher yields. U.S. dry bean production is already characterized by large-scale farms, and recent yields exhibit a modest upward trend.

On the demand side, per capita disappearance of dry beans in Mexico is declining over the long term as consumers shift away from traditional staple foods. U.S. per capita consumption is relatively stable but only at about a fourth of Mexico's average level, even though dry beans are a low-fat source of important nutrients.

* This is an edited, reformatted and augmented version of the United States Department of Agriculture publication, VGS-341-01, dated December 2010.

Keywords: Dry beans, Mexico, United States, NAFTA, North American Free Trade Agreement

INTRODUCTION

Participants in U.S.-Mexico dry bean trade are steadily gaining experience with how the North American market operates under conditions of regional free trade. The last transitional tariff-rate quotas (TRQs)[1] on U.S.-Mexico (and Canada-Mexico) dry bean trade were removed at the start of 2008, in accordance with the North American Free Trade Agreement (NAFTA). As a result, the outlook for the U.S. and Mexican dry bean sectors is now shaped by more fundamental forces of supply and demand, rather than anticipated reductions in regional trade barriers.

In this report, the authors examine the significance of dry bean trade to the NAFTA countries; provide a detailed understanding of supply, demand, and policy in the U.S. and Mexican dry bean sectors; and consider the outlook for these industries. Special attention is given to the changing importance of dry beans within Mexican society. Dry beans continue to be an important source of protein, carbohydrates, and other nutrients in many Mexican households, particularly those of modest means. Nevertheless, consumption has declined over the past two decades in both aggregate and per capita terms, and this trend is likely to continue. Moreover, while producers working less than 5 hectares of land account for roughly half of Mexico's dry bean farmers, the sector is gradually becoming more consolidated, smaller in terms of area planted, and more efficient in terms of yields.

BEAN CLASSES AND VARIETIES COVERED BY NAFTA'S LAST SET OF IMPORT RESTRICTIONS

Mexico and the United States produce a diversity of dry bean varieties, not all of which were covered by the transitional import restrictions that expired in 2008 as part of NAFTA. The production statistics collected by USDA's National Agricultural Statistics Service (NASS) cover the following commercial classes of dry beans: large lima, baby lima, navy, Great Northern, small white, pinto, light red kidney, dark red kidney, pink, small red, cranberry, black, blackeye, small chickpeas, and large chickpeas. Each class includes multiple varieties; for instance, the United States cultivates over 15 varieties of black beans. NAFTA's TRQs for U.S. and Canadian dry bean exports to Mexico for the period 1994-2007 (tariff line 0713.33.02 in Mexico's NAFTA tariff schedule) covered beans belonging to the species *Phaseolus* vulgaris, or "common" beans.

The dry bean classes covered by these TRQs—pinto, navy, red kidney, black, Great Northern, small red (excluding Adzuki), pink, cranberry, and small white—account for about 90 percent of U.S. production. Within this framework, the leading classes of dry common beans produced in the United States are pinto (48 percent), navy (17 percent), black (13 percent), and red kidney (8 percent) (fig. 1).

Given the focus of NAFTA's restrictions and longstanding interest in U.S.- Mexico trade in dry common beans, the authors primarily examined dry common beans, and the statistics presented in this report for U.S. dry bean exports, imports, and production cover only common classes and varieties of dry beans. Appendix table 1 identifies the tariff lines considered as dry common beans for the purposes of this report.[2]

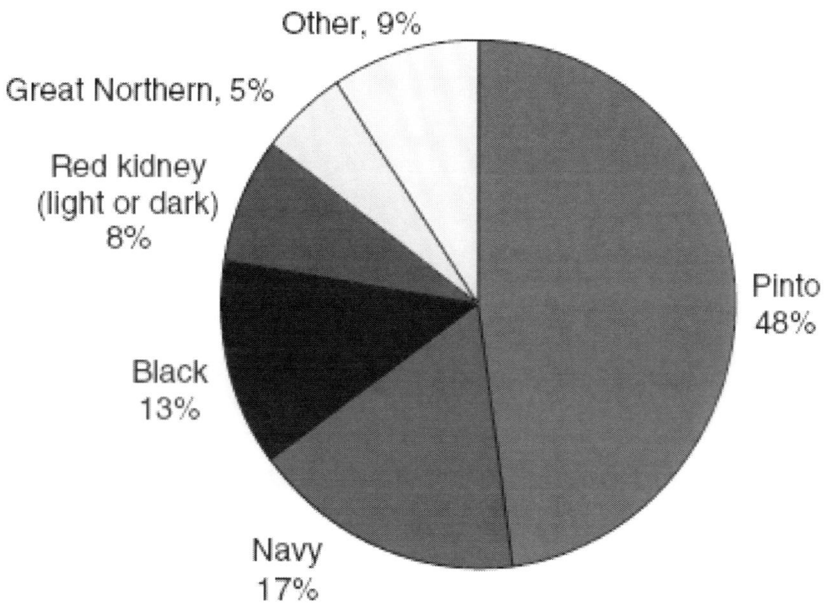

Source: USDA, National Agricultural Statistics Service, Crop Production, January 2009.

Figure 1. U.S. dry bean production, by variety, 2007-09.

Mexico produces about 50 different varieties of dry beans, and these may be divided into four major groups (fig. 2). Black (*negros*) is the group with the largest share of total consumption (38 percent), followed by pintos (26 percent), pink (*rosas*, also referred to as *claros* or light-colored beans, 20 percent), and yellow (*amarillos*, also referred to as *azufrados* or sulfur-colored, 16 percent). The most commonly produced varieties of pink beans in Mexico are *Flor de Mayo* and *Flor de Junio*. Mexico's dry bean exports to the United States include some varieties that once were rarely produced in the United States, such as the yellow variety known as *peruano* or *mayocoba* (Wingett, 2006). U.S. and Mexican trade statistics are not sufficiently detailed to specify the actual volumes of this trade, however.[3]

Some producers in Mexico are adopting different varieties of dry beans. In the State of Durango, for instance, the Government and the dry bean sector are encouraging producers with a medium to high level of potential productivity to plant certified varieties better suited for precipitation conditions in Durango (Guzmán, 2010). Production in the State of Zacatecas has shifted toward pinto beans and away from the traditional light-colored varieties of Flor de Mayo and Flor de Junio. Pintos are projected to account for 23 percent of Zacatecas's dry bean production in 2010, compared with 3 percent in 2002 (Vallejo Díaz, 2010). Since 2004, the Idaho Bean Commission has worked with Mexican producers to test varieties suited for specific Mexican growing regions—first in the State of Sinaloa and later in the States of Chihuahua and Zacatecas—with the intent of increasing sales of certified seed to Mexico (Idaho State Department of Agriculture, 2006; Stewart-Williams, 2009; and Wilkins, 2009).

THE IMPORTANCE OF DRY BEAN TRADE TO NAFTA COUNTRIES

Dry bean trade is a key component of sales for U.S. and Canadian producers and an important source of supply for Mexican consumers (table 1). Mexico is the destination for about 35 percent of U.S. dry bean exports and 10 percent of U.S. dry bean production, while Canada receives about 10 percent of U.S. dry bean exports and 3 percent of U.S. production. For Mexico, imports from the United States account for about 10 percent of national dry bean supply (calculated as imports plus domestic production), stabilizing consumption when downturns in production occur. For Canada, exports to the

United States and Mexico combined to account for about 14 percent of dry bean production.[4]

According to Mexican trade statistics, the United States is Mexico's principal foreign supplier of dry beans, with a 95-percent share of Mexico's total dry bean imports. U.S. trade data contain detailed information about the classes and varieties of dry beans exported to Mexico, although there is some question as to whether these details have been consistently and correctly identified over time. According to available data, black beans were the leading dry bean class exported to Mexico during U.S. marketing years[5] 2006/07 to 2008/09, accounting for 52 percent of U.S. dry bean exports to Mexico (fig. 3). Pinto beans accounted for 32 percent.

The architects of NAFTA recognized the possible sensitivity of Mexico's dry bean producers to import competition and provided for the liberalization of Mexican dry common bean imports from the United States and Canada over a period of 14 years (1994-2007), the longest transitional period specified by the agreement for agricultural products. Other varieties of U.S. and Canadian dry beans, such as garbanzo, lima, blackeye, and Adzuki, have enjoyed duty-free access to the Mexican market since 1994, when NAFTA initially took effect.[6]

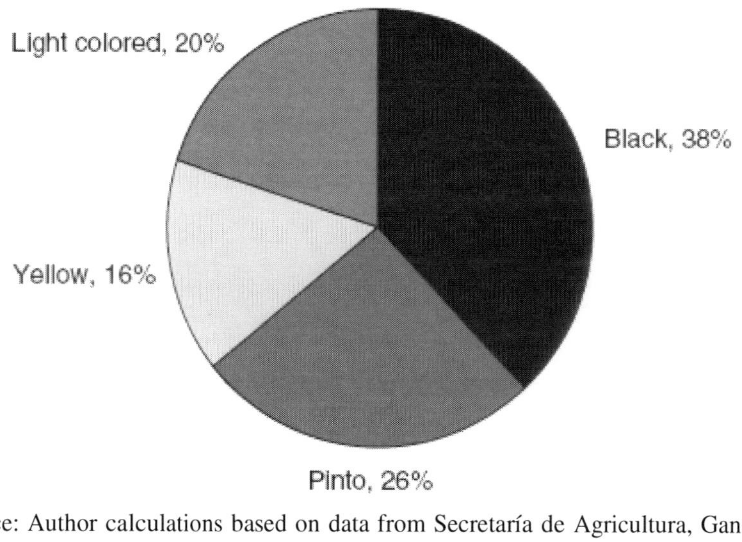

Source: Author calculations based on data from Secretaría de Agricultura, Ganadería, Desarrollo Rural, Pesca, y Alimentación, Servicio de Información Agroalimentaria y Pesquera (SAGARPA/SIAP).

Figure 2. Distribution of Mexican dry bean consumption, by variety, 2008.

Table 1. Overview of North American dry bean production and trade: Annual averages, U.S. marketing years 2006/07 to 2008/09

					Exports to:					Imports from:		
Reporting country	Production		World	Canada	Mexico	U.S.	ROW	World	Canada	Mexico	U.S.	ROW
United States	1,164	*	330	33	116	--	181	72	34	7	--	31
Mexico	1,027	**	20	***	--	17	3	116	9	--	107	***
Canada	305	*	216	--	7	37	172	34	--	***	29	5
Total, North America	2,496		566	33	123	54	356	222	43	7	136	36

Source: Author calculations based on data from Secretaría de Agricultura, Ganadería, Desarrollo Rural, Pesca, y Alimentación, Servicio de Información Agroalimentaria y Pesquera (SAGARPA/SIAP, 2010a) (Mexican production); Secretaría de Economia, as cited by Global Trade Information Services, Inc. (2010) (Mexican trade data); Statistics Canada, as cited by Agriculture and Agri-food Canada (2009) (Canadian production); Statistics Canada, as cited by Global Trade Information Services (2010) (Canadian trade data); USDA, National Agricultural Statistics Service, Crop Production, January issues (2009-10) (U.S. production); U.S. Department of Commerce, Census Bureau, Foreign Trade Statistics, as cited by USDA, Foreign Agricultural Service (2010) (U.S. trade data).

ROW = Rest of world.

-- = Data not applicable.

* = Annual average for 2006-08.

** = Annual average for Mexico's 2006-08 agricultural years.

*** = Less than 500 metric tons.

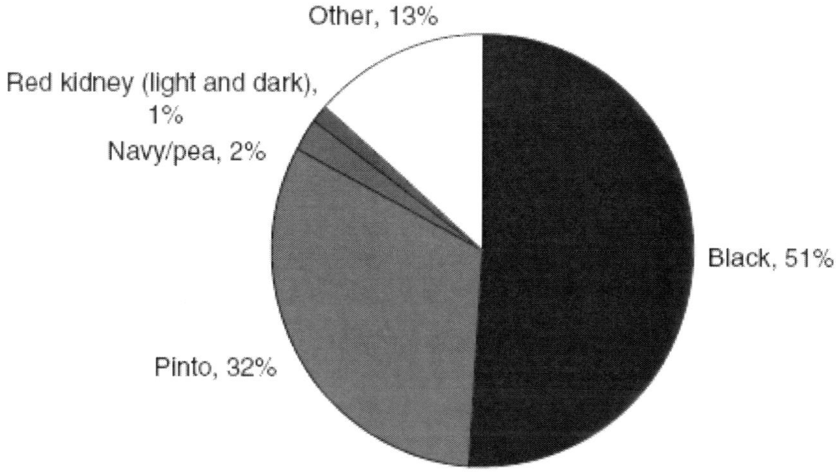

Source: Author calculations based on data from U.S. Department of Commerce, Census Bureau, Foreign Trade Statistics, as cited by USDA, Foreign Agricultural Service (2010).

Figure 3. U.S. dry bean exports to Mexico, by class, marketing years 2006/07-2008/09.

NAFTA has facilitated an important change in the U.S.-Mexico trading relationship for dry beans. Before NAFTA, the Mexican Government tightly regulated dry bean imports from the United States through a system of import licenses. As a result, imports fluctuated widely from year to year, with little to no trade when Mexican production was ample and large volumes of trade when Mexican production was limited (fig. 4). Since NAFTA's implementation, U.S. dry bean exports to Mexico have become far more consistent, averaging 116,000 metric tons (mt) per year during 2006/07 to 2008/09. Nevertheless, trade continues to fluctuate in response to weather conditions in either country.

Trade liberalization also has facilitated the participation of Mexican and Canadian producers in the U.S. market, although current U.S. most-favored-nation (MFN)[7] import tariffs on dry beans are quite small, with *ad valorem* values of 2 percent or less. Imports from Mexico averaged about 7,000 mt per year during 2006/07-2008/09, compared with less than 1,000 mt per year during the 3 marketing years immediately prior to NAFTA (1990/91-1992/93). Imports from Canada also have increased, averaging 34,000 mt per year during 2006/07-2008/09. Prior to the implementation of the CanadaU.S. Free Trade Agreement (CUSTA)[8] in 1989, these imports had never exceeded 15,000 mt per year.

Exports to countries outside NAFTA—the rest of the world (ROW) in table 1—are a significant outlet for U.S. and Canadian dry beans, but not for Mexican dry beans. During 2006/07-2008/09, such exports accounted for 16 percent of U.S. production and 56 percent of Canadian production. In markets outside the NAFTA region, U.S. producers have faced heightened competition from China, Burma, and Canada, each of which increased its annual dry bean exports by 100,000 mt or more between 1991-93 and 2005-07 (appendix table 3).[9] In this context, U.S. dry bean exports to the ROW declined from 256,000 mt in 1998/99 to 132,000 mt in 2004/05, a decrease that corresponded with about 9 percent of U.S. production in 1998. Much of this decrease is linked to declining exports of navy beans to the European Union (EU). Canada, in particular, has become much more competitive on a cost and a volume basis, vying with U.S. products in both the EU and the United States.

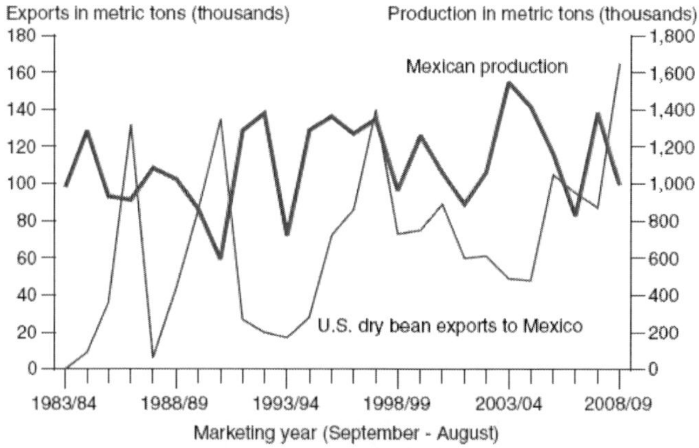

Source: Author calculations based on data from USDA, Foreign Agricultural Service (2010); Secretaría de Agricultura, Ganadería, Desarrollo Rural, Pesca, y Alimentación, Servicio de Información Agroalimentaria y Pesquera (SAGARPA /SIAP, 2010a).

Notes: The production statistics in this figure correspond to Mexico's agricultural years. Mexico's agricultural year is divided into two production cycles: fall/winter and spring/summer. To compare U.S. exports with Mexican production, we matched U.S. marketing years and Mexican agricultural years so that the starting year of the marketing year is the same number that denotes the agricultural year. For instance, the U.S. marketing year 2008/09 is matched with Mexico's 2008 agricultural year. This allows us to compare the quantities of U.S. and Mexicandry beans that are on the market at roughly the same time.

Figure 4. U.S. dry bean exports to Mexico and Mexican production of dry beans, 1983/84 to 2008/09.

Table 2. Selected price data for the U.S. and Mexican dry bean sectors

	U.S. dry bean marketing year				
	2005/06	2006/07	2007/08	2008/09	Change, 2008/09 versus 2007/08
Consumer price indices:	Indices				Percent
United States, overall					
(1982-84 = 100)	200	205	214	214	0.2
United States, dry					
beans (1997 = 100)	119	127	147	175	19.2
Mexico, overall	117	121	127	134	5.9
Mexico, dry beans					
(2Q 2002 = 100)	99	102	111	155	39.5
Exchange rate:		*Pesos per dollar*			*Percent*
Nominal	10.84	10.93	10.64	13.25	24.5
Real (2005)	10.87	10.78	10.48	12.33	17.7
Pinto beans:		*Dollars per mt*			
Bids to growers	307	466	617	593	-3.8
Dealer prices	517	651	852	901	5.6
Unit value, exports to Mexico	560	670	785	880	12.2
		Pesos per mt			
Unit value, exports to Mexico	6,070	7,327	8,326	11,640	39.8
Wholesale price, imports, Mexico City	10,243	9,574	11,723	15,321	30.7
Wholesale price, domestic product, Mexico City	9,365	8,191	10,150	14,556	43.4
Black beans:		*Dollars per mt*			
Bids to growers,	461	528	685	741	8.1
Michigan					
Dealer prices, Michigan	651	682	958	1,029	7.4
Unit value, imports from	583	656	858	987	15.1
U.S.					
		Pesos per mt			
Unit value, imports from	6,329	7,169	9,116	13,098	43.7
U.S.					
Wholesale price, imports, Mexico City	11,289	11,396	12,995	16,274	25.2
Wholesale price, domestic product					
(Veracruz), Mexico City	8,758	8,438	8,490	14,018	65.1

Source: Author calculations based on data from Banco de Mexico (2010) (Mexico CPI); Secretaría de Economía (2010) (Mexico wholesale prices); USDA, Foreign Agricultural Service (2010) (unit values); USDA, Agricultural Marketing Service (2005-09) (bids to growers and dealer prices); USDA, Economic Research Service (2010b, c) (exchange rates); and U.S. Department of Labor, Bureau of Labor Statistics (2009) (U.S. CPI).

mt – Metric ton.

Note: Mexico City prices are for the Central de Abasto, Iztapalapa.

A strengthening of the real value of the U.S. dollar vis-à-vis its Canadian counterpart during 2002/03 to 2006/07 boosted the competitiveness of Canadian products. Over the past 4 years, however, the area planted with dry beans in Canada has experienced successive annual declines, driven partly by reductions in white and navy bean area (Lucier, Dettman, and Da Pra, 2009: p. 19) and a strengthening of the Canadian dollar in 2007/08 and 2008/09. These developments have facilitated a modest rebound in U.S. dry bean exports to the ROW. These exports reached 233,000 mt in 2008/09, their highest level in 8 years.

PRICE DATA REVEAL SIMILAR MOVEMENTS IN TWO MARKETS

A close relationship between the prices of a specific good in two geographically distinct locations is viewed by many economists as a defining characteristic of integrated markets (Vollrath, 2003). While a formal analysis of price integration is beyond the scope of this report, the available data suggest some similarities in U.S. and Mexican dry bean price movements during U.S. marketing year 2008/09 (table 2). Practically all of the consumer, producer, and wholesale prices for dry beans listed in table 2 increased between 2007/08 and 2008/09, reflecting the broader phenomenon of higher prices for many agricultural commodities during that period. In both countries, the price paid by urban consumers for dry beans increased at a much faster rate than the overall consumer price index (CPI) in 2008/09. For Mexican consumers, however, the increase in the dry bean CPI was proportionately larger than for U.S. consumers; 39.5 percent in Mexico, compared with 19.2 percent in the United States.

For the two leading classes exported to Mexico (pinto and black), U.S. bids to growers and dealer prices increased for black beans but decreased for pinto beans between 2007/08 and 2008/09.[10]

In Mexican wholesale markets, as exemplified by the Central de Abasto in the Iztapalapa delegation of Mexico City, the prices of domestically grown pinto and black beans increased at a faster rate in 2008/09 than corresponding prices of imported product. At the same time, the increases in the wholesale prices of imported products (30.7 percent for pinto beans and 25.2 percent for black beans) were substantially less than the 39.5-percent increase in the Mexican CPI for dry beans. Thus, the availability of U.S. dry beans in the

Mexican market may have tempered the impact of higher prices of domestically produced dry beans on Mexican consumers.[11]

GEOGRAPHY AND FARM SIZE DIFFERENTIATE U.S. AND MEXICAN DRY BEAN SECTORS

Mexico and the United States are similar dry bean producers in terms of total output. Mexico ranks sixth among the world's leading dry bean producers, with average annual output of 1.05 million metric tons (mmt) during agricultural years 2007-09;12 the United States ranks fifth, with average annual output of 1.16 mmt during 2007-09.13 At the same time, dry beans account for a small portion of the total value of crop production in the two countries: about 3 percent in Mexico and 0.5 percent in the United States (SAGARPA/ SIAP, 2010a; and USDA/NASS, 2010a). But the geographic and socioeconomic conditions under which dry beans are cultivated in the two countries are quite different in many respects, and these differences are reflected in the production statistics.

Climate's Impact on Yields

Figure 5 indicates the location of U.S. and Mexican dry bean production, using U.S. county-level data from the 2007 Agricultural Census and Mexican municipal-level data for agricultural years 2006-08 (see box, "Where Are Dry Beans Grown?"). U.S. production takes place either in regions with a humid continental climate—marked by cold winters and warm summers—or in semiarid places where irrigation is available. In contrast, Mexican production tends to take place either in a semiarid climate—often without the benefit of irrigation—or in a tropical savannah climate featuring a dry season and a wet season and the possibility of severe weather in either season. Some Mexican producers use irrigation to compensate for the dry climate, but most lack access to large amounts of irrigated water. Rainfed lands accounted for 86 percent of Mexico's area harvested with dry beans and 69 percent of Mexican dry bean production during agricultural years 2007-09. As a result of these cross-country differences in climate and availability of irrigation, U.S. dry bean yields are substantially higher than Mexico's. U.S. yields averaged 1.98

mt per hectare during 2007-09, while Mexican yields averaged 0.76 mt per hectare during agricultural years 2007-09 (fig. 6).

Lower yields are important to the Mexican dry bean sector for several reasons. First, they contribute to the persistent fragmentation of the sector by reducing the economic gains from consolidating smaller plots and spreading the fixed costs of production across a larger quantity of output. Second, they represent an agronomic challenge that is not easily surmounted. Irrigated lands generate roughly 30 percent of Mexico's dry bean production, and the costs of this irrigation are often borne by persons outside the dry bean sector. The electricity used by farmers to pump irrigation water is subsidized, and the charges for water often do not reflect its true economic value.[14] In the absence of irrigation, the prospect for higher yields lies squarely in the development of technological innovations suited for rainfed lands, such as improved seed varieties.

Where Are Dry Beans Grown?

Dry bean production in Mexico and the United States is concentrated in certain regions. In Mexico, three regions are responsible for about 80 percent of the country's dry bean production: (1) North-Central Mexico (the States of Zacatecas, Durango, Chihuahua, San Luis Potosí, and Guanajuato); (2) the Northern Pacific Coast (the States of Sinaloa and Nayarit); and (3) the State of Chiapas. North-Central Mexico accounted for 52 percent of Mexico's dry bean production during 2007-09. Much of this region's production takes place under semi-arid conditions, although in some parts of the region, dry bean production takes place in a variety of climates (Vallejo Díaz, 2010). For instance, in Zacatecas, dry bean production occurs in arid, semi-arid, and subtropical areas. Roughly 6 percent of North-Central Mexico's bean area is irrigated, and nearly all (99 percent) of the region's bean production occurs during the spring-summer agricultural cycle. The Northern Pacific Coast region accounted for 21 percent of Mexico's dry bean production during 2007-09. This region's climate ranges from desert and semi-arid in Sinaloa to tropical savannahs in Nayarit. Production in the Northern Pacific Coast region takes place almost entirely during the fall-winter agricultural cycle, and about two-thirds of the region's bean area is irrigated. Chiapas, located primarily in a tropical savannah climate, accounted for 7 percent of Mexican production during 2007-09. Roughly 62 percent of Chiapas's production occurs during the spring-summer cycle, and very little of its production is irrigated.

Similarly, three regions account for about three-fourths of U.S. dry bean production: (1) the Northern Great Plains (North Dakota and Minnesota); (2) the States along the Platte River (Nebraska, Colorado, and Wyoming), and (3) Michigan. The Northern Great Plains region accounted for 53 percent of U.S. production during 2007-09.

North Dakota, the top producing State, grows several classes of dry beans—including pinto, navy, and black—while navy and dark red kidney beans are the main classes produced in Minnesota. According to the 2007 Agricultural Census, only 2 percent of North Dakota's area harvested with dry beans is irrigated, compared with 18 percent in Minnesota (USDA/NASS, 2010b).

Production in Nebraska, Colorado, and Wyoming takes place primarily along the western portion of the Platte River, including both the North and the South Platte Rivers. This region, which accounted for 17 percent of U.S. dry bean production during 2007-09, primarily produces pinto and Great Northern beans.

In contrast to other dry bean producing regions in the United States, Platte River States account for a relatively high proportion of irrigated production—90 percent of area harvested.

The third region, Michigan, accounted for 14 percent of U.S. dry bean production during 2007-09. Michigan is second in U.S. navy bean production and the Nation's leading producer of black beans. Producing about 57 percent of U.S. black beans, Michigan is also the leading U.S. supplier of black beans to Mexico. Very little of Michigan's dry bean production (4 percent of area harvested) is irrigated.

U.S. dry bean yields differ substantially by commercial class. During 2007- 09, the average yield ranged from 1.72 mt per hectare for cranberry beans to 2.68 mt per hectare for small white beans (appendix table 6). This wide range in yields is due to the influence of irrigation and region of production on yields.

Black beans are generally produced on nonirrigated lands in some States, such as Michigan and North Dakota, while irrigated lands in Idaho and Oregon account for a large portion of U.S. production of small white beans. Classes that are commonly irrigated (primarily because of where they are grown) include Great Northern, small white, pink, and small red beans.

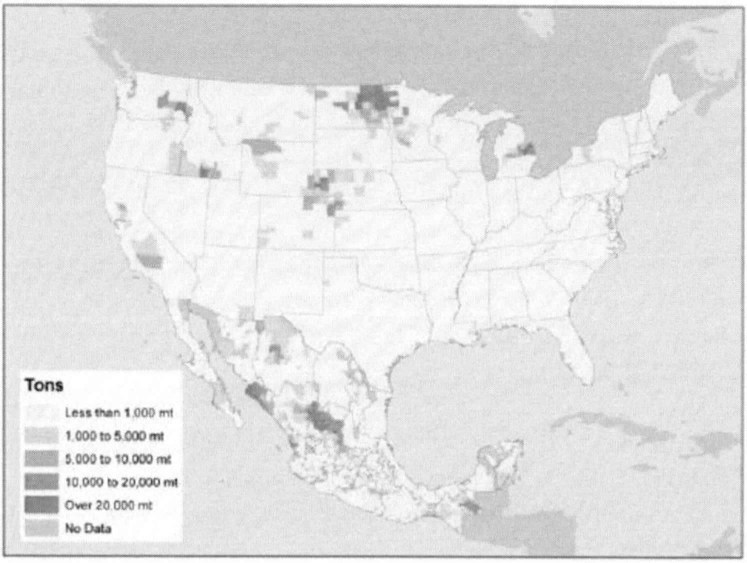

Source: Secretaría de Agricultura, Ganadería, Desarrollo Rural, Pesca, y Alimentación, Servicio de Información Agroalimentaria y Pesquera (SAGARPA/SIAP, 2010a); USDA, National Agricultural Statistics Service, 2007 Census of Agriculture.

mt=Metric ton.

Note: The Mexican production statistics in this report correspond to Mexico's agricultural years, while the U.S. production statistics correspond to calendar years. Mexico's agricultural year is divided into two production cycles: fall/winter and spring/summer. U.S. county-level data are for 2007; Mexican municipal data are annual averages from agricultural years 2006-08.

Figure 5. Locations of dry bean production in the United States and Mexico.

Units of Production

Another key difference between Mexico and the United States is the scale of a typical dry bean operation. A typical U.S. dry bean producer works about five times as much land as a typical Mexican producer who is focused primarily on the commercial market—about 100 hectares in the United States, compared with roughly 20 hectares in Mexico.[15] Moreover, Mexico has a large number of dry bean farmers who operate on an even smaller scale and thus have limited marketable surpluses.

According to Mexico's 2007 Agricultural Census (Instituto Nacional de Estadística y Geografía (INEGI), 2009b,c), about 622,000 farms grow dry beans in Mexico, with an average area planted of 3.1 hectares and an average area harvested of 2.8 hectares.

Over the past two decades, dry bean production in the United States and Mexico has become concentrated on a smaller number of farms. In the United States, the number of farms producing dry beans dropped from about 15,900 to 6,200 between 1987 and 2007, while the area of dry beans harvested per farm climbed from 42 hectares to 94 hectares (table 3). In Mexico, the decrease in the number of farms growing dry beans has coincided with a decrease in the area planted of this crop. The average size of a dry bean parcel in Mexico's spring-summer agricultural cycle increased 9.4 percent between 1991 and 2007, while the average parcel size in the fall-winter cycle decreased 7.6 percent (table 4).

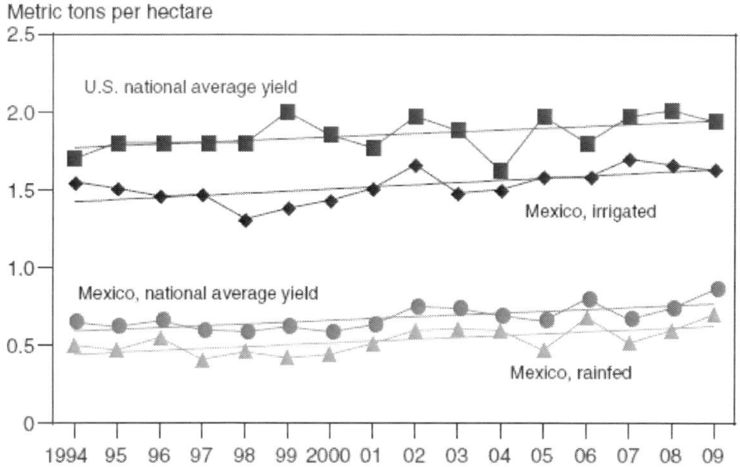

Notes: The straight lines represent the linear trend lines for each data series. The Mexican production statistics in this report correspond to Mexico's agricultural years, while the U.S. production statistics correspond to calendar years. Mexico's agricultural year is divided into two production cycles: fall/winter and spring/summer.

Source: Secretaría de Agricultura, Ganadería, Desarrollo Rural, Pesca, y Alimentación, Servicio de Información Agroalimentaria y Pesquera (SAGARPA/SIAP, 2010a); USDA, National Agricultural Statistics Service, Crop Production, January issues (1997-2010).

Figure 6. U.S. and Mexican dry bean yields have trended upward during the NAFTA period.

An analysis of these production trends at the State level reveals that a consolidation of the Mexican dry bean sector has taken place in two of the leading producing States.

In Zacatecas, the average size of a parcel cultivated with dry beans increased from 7.7 to 9.5 hectares between 1991 and 2007, while in Sinaloa parcels increased from 8.9 to 12.2 hectares. At the same time, the State of Chiapas continues to be one of Mexico's larger dry bean producers, even though its average dry bean parcel size remains around 1 hectare, well below the national average.

Data from Mexico's main direct-payment program suggest that roughly half of Mexico's dry bean producers devote 5 hectares of land or less to dry beans. Producers of this scale tend to focus on local markets, including the subsistence of their own households, and consolidating their supply with other producers.

About 20 percent of Mexico's dry bean production is used for home or local consumption, while 70 percent enters regional and national commercial circuits (fig. 7).

Among the latter groups, the main destinations are wholesale markets and packing houses. While many of Mexico's smaller dry bean producers focus on local markets, the producing regions for a particular variety are often quite distant from the destination markets for that variety (fig. 8); a feature that some observers view as a marketing and competitiveness problem (at least for some producers) due to high transportation and storage costs (World Bank, Agriculture and Rural Development Team, Sustainable Agriculture Department, Latin America and the Caribbean Region; Instituto Mexicano para la Competividad, 2007: pp. 38, 56-9).

The extent to which small-scale production continues in Mexico's dry bean sector will have a marked impact on the sector's future. In 2007, the average age of a Mexican farmer (for all crops) was 51 years, compared with 57 years in the United States, and the retirement of older, small-scale farmers without their full replacement by a younger generation will necessarily result in a more consolidated dry bean sector where producers that are more closely oriented to the commercial market play a larger role.

At the same time, Mexico's traditional diet based on white corn and dry beans was linked closely to small-scale agricultural production, so the retirement of small-scale dry bean producers could also accompany further reductions in per capita dry bean consumption in Mexico.

Table 3. U.S. dry bean sector: Number of farms and area harvested, 1987-2007

Census year	Number of farms	Area harvested	Average dry bean area per farm
		Hectares	
2007	6,236	589,040	94
2002	8,647	684,637	79
1997	11,348	700,874	62
1992	13,201	626,763	47
1987	15,914	675,864	42

Source: USDA, National Agricultural Statistics Service (2009a, 2004, 1999, 1994).

Table 4. Mexican dry bean sector: Number of farms and area planted, 1991 and 2007

Agricultural cycle	Units of production	Area planted with dry beans	Average parcel size
	Number	Hectares	
Fall-winter 1991	155,856	433,669	2.78
Spring-summer 1991	814,095	2,384,856	2.93
Fall-winter 2007	95,854	246,366	2.57
Spring-summer 2007	526,410	1,686,680	3.20
2007 compared with 1991:		Percentage change	
Fall-winter	-38.5	-43.2	-7.6
Spring-summer	-35.3	-29.3	9.4

Source: Author calculations based on data from Instituto Nacional de Estadística y Geografía (INEGI), VII Censo Agrícola, Ganadero y Forestal, 1991, and VIII Censo Agrícola, Ganadero y Forestal, 2007.

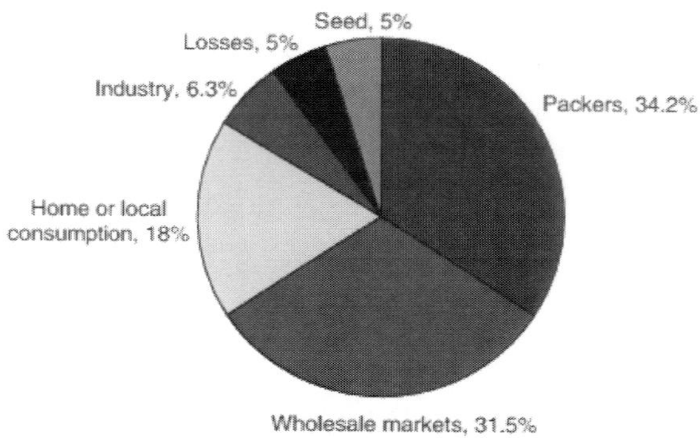

Source: Author calculations based on data from Plan Rector Sistema Producto Frijol 2008.

Figure 7. Destinations for Mexican dry bean production, 2008.

Smaller, More Productive Dry Bean Sectors Have Emerged in Both Countries

Dry bean production varies considerably from year to year as a result of changes in area planted, area harvested, and yields. Nevertheless, several long-term trends are visible in Mexican and U.S. dry bean production, broadly characterized by less area devoted to the crop and higher yields. In Mexico, area planted with dry beans has trended downward, from 2.2 million hectares in 2002 to 1.7 million hectares in 2009 (fig. 10), as farmers reallocated some land to alternative crops and/or livestock production—sometimes in response to market price signals and sometimes with governmental support. Area harvested also has trended downward, from 2.1 million hectares in 2002 to 1.2 million hectares in 2009. Rainfed lands account for about 80 percent of these decreases. Programs that encourage dry bean farmers working marginal lands to shift toward other agricultural activities may have reduced the loss rate (the percentage difference between area harvested and area planted)[16] in Mexico's dry bean sector, although the lower loss rate may also be the result of favorable weather conditions.

The median loss rate was 10 percent during 2000- 09, compared with 13 percent during 1990-99. Dry beans are extremely sensitive to excessive humidity, frost, and drought, and this vulnerability leads to very high loss rates in certain years. For example, the loss rate reached 28 percent in 2005 and 2009, years in which droughts negatively affected dry bean production in some parts of Mexico.

Mexico's dry bean yields on irrigated lands have experienced a fairly steady, upward trend since 1980, while yields on rainfed lands have shown only signs of a sustained improvement over the last decade (see fig. 6). Still, national yields have grown at a sufficient rate so that, together with the decreased loss rate, they offset the decrease in area harvested. As a result, Mexican production has fluctuated around the level of 1.2 mmt over the past 15 years, even though many of the year-to-year changes in output have been dramatic (fig. 9).

In the United States, the dry bean sector is much smaller than it was a decade ago (fig. 10). During 2007-09, production averaged 1.0 mmt per year, compared with 1.2 mmt during 1996-98. During these same two periods, area harvested averaged 526,000 hectares and 686,000 hectares, respectively. U.S. dry bean production has risen and fallen over the past decade with the competitiveness of navy bean exports and domestic consumer preferences for dry beans. As discussed in the trade section, U.S. navy bean exports experienced a modest rebound during marketing years 2005/06 to 2008/09, and this improved export performance was reflected in higher production of both navy beans and dry beans. Production of dry bean classes other than navy has fluctuated from one year to the next, but the long-term trend is stable.

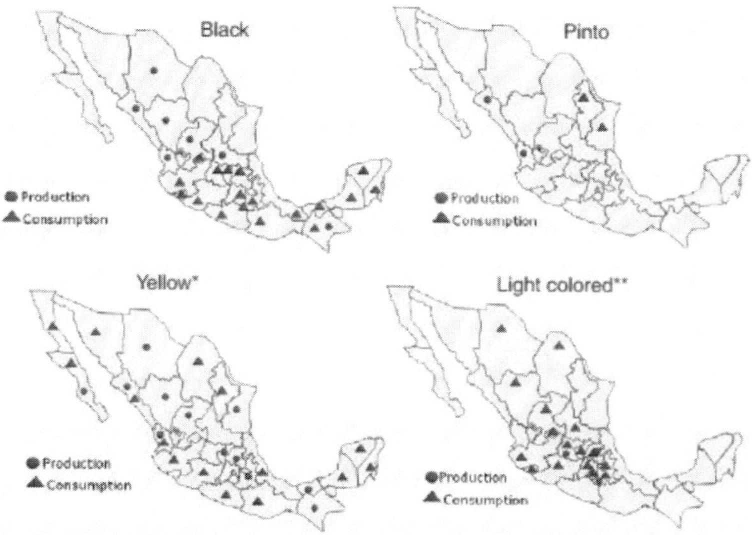

Source: Author presentation of data from Secretaría de Agricultura, Ganadería, Desarrollo Rural, Pesca, y Alimentación, Apoyos y Servicios a la Comercialización Agropecuaria (SAGARPA/ASERCA), General Coordinator of Commercialization.

*=Includes principally the varieties known as azufrado, garbancillo zarco, mayocoba, and peruano.

**=Includes principally the varieties known as Flor de Mayo and Flor de Junio.

Note: The circles identify the major States where a particular variety is produced, while the triangles identify States where a particular variety tends to be consumed.

Figure 8. Production and consumption of dry beans, by variety and region.

The national loss rate in the U.S. dry bean sector is usually several percentage points less than that found in Mexico. During 2005-09, the median national loss rate in the United States was 5 percent. U.S. dry bean producers are fortunate to avoid the high loss rates that their Mexican counterparts occasionally see. Over the past two decades, the highest national U.S. loss rate was 13 percent for 1993 and 2001.

Compared with leading U.S. field crops, such as corn and soybeans, less research has been done to develop new dry bean varieties over the past several decades. Nevertheless, U.S. dry bean yields (for all varieties) have been on a gradual upward trend over the last quarter century, which is due partly to favorable weather (see fig. 6). When considering individual varieties and major producing States, there are only isolated instances where there was a strong tendency for yields to rise over the past 15 years. A key example is pinto beans in Nebraska, where much of dry bean production is irrigated.

Mexico's Dry Bean Policies Aim to Increase Sector Productivity

The future of the dry bean sector is a vital issue to many Mexicans, partly because of the large number of small-scale producers, but also because the commodity takes on great economic importance in States, such as Zacatecas, Durango, Chihuahua, San Luis Potosí, Sinaloa, and Nayarit, that have substantial surpluses in dry bean production and are seeking additional market opportunities for their output.

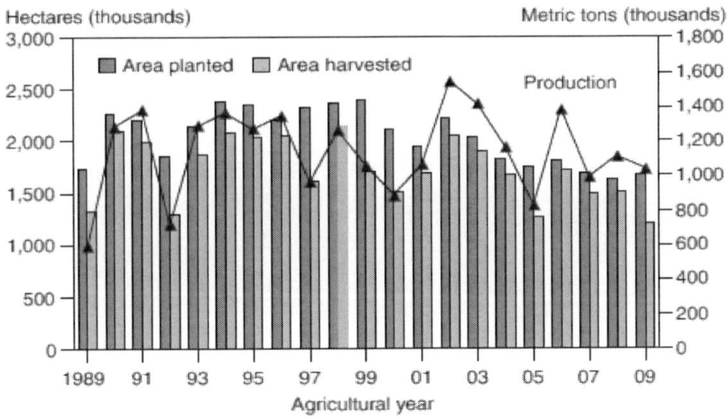

Source: Secretaría de Agricultura, Ganadería, Desarrollo Rural, Pesca, y Alimentación, Servicio de Información Agroalimentaria y Pesquera (SAGARPA/SIAP, 2010a).

Notes: The Mexican production statistics in this report correspond to Mexico's agricultural years, while the U.S. production statistics correspond to calendar years. Mexico's agricultural year is divided into two production cycles: fall/winter and spring/summer.

Figure 9. Mexican dry bean area and production, 1989-2009.

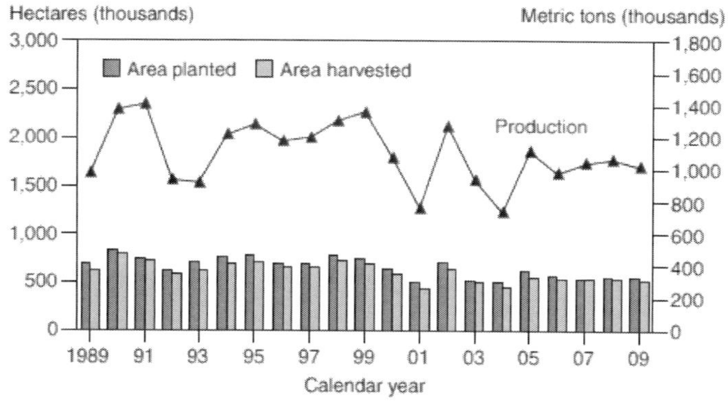

Source: USDA, National Agricultural Statistics Service, Crop Production, January issues (1990-2010).

Notes: The Mexican production statistics in this report correspond to Mexico's agricultural years, while the U.S. production statistics correspond to calendar years. Mexico's agricultural year is divided into two production cycles: fall/winter and spring/summer.

Figure 10. U.S. dry bean area and production, 1989-2009.

In light of the perceived importance of the dry bean sector, the Mexican Government operates a number of programs designed to increase its productivity, focusing on direct income support, commercialization, access to credit, adoption of superior technologies, and the organization of productive chains and product systems.

The Program of Direct Support for the Countryside (PROCAMPO—Programa de Apoyos Directos para el Campo) is Mexico's main program that provides direct income support to agricultural producers. PROCAMPO was developed in 1993 to compensate farmers for possible losses they might experience as a result of NAFTA and to improve their competitiveness in the face of other countries' farm supports. Under PROCAMPO, any producer who cultivates a legal crop on eligible land or uses that land for livestock or forestry production or some ecological project can receive payments, which are made on a per hectare basis. Eligible land is defined as that which had been cultivated with corn, dry beans, wheat, rice, sorghum, soybeans, cotton, safflower, soybeans, or barley in any of the three agricultural cycles prior to August 1993.

Since its inception, PROCAMPO has been modified several times to make the program more progressive in terms of income distribution. As a result of these modifications, PROCAMPO now gives preferential treatment to smaller producers during the spring-summer cycle so that they may receive larger transfers per hectare. According to operational rules published on April 8, 2009, the standard PROCAMPO payment rate is 963 pesos (about U.S. $77) per hectare for both the fall-winter and spring-summer agricultural cycles, while the spring-summer payment rate equals 1,300 pesos (U.S. $104) for those producers with less than 5 hectares of eligible land. Until December 2008, the opportunity also existed for producers enrolled in PROCAMPO to access their payments in advance to receive financing for productive projects from banking institutions. This program, known as PROCAMPO Capitaliza, also gave preference to farmers with less than 5 hectares of eligible land.

Mexico's dry bean farmers also count on a series of Government programs designed to facilitate the commercialization of their product and to avoid the negative price effects of excess production or market downturns. Specific commercialization programs for dry beans in 2009 included:

1. *Support for Crop Storage:* Partially or totally covers the costs of storage and financing for a period of 6 months and, in some cases, up to 9 months.
2. *Support for the Transport of Bean Harvests:* Partially or totally covers the shipping costs associated with transporting beans from the production region to consumption region.
3. *Support for Value Added:* Partially or totally covers the costs incurred from improving the quality and presentation of the product for sale.

To remedy excess production, the Mexican Government has established mechanisms for surplus management, utilizing pledge schemes to avoid price volatility. These schemes are oriented toward producer organizations in States with surpluses and/or marketing problems and provide assistance for as much as 25 percent of the total harvest volume.

Access to credit is one of the main challenges facing Mexico's dry bean sector. This problem arises partly from the inherent risks of agricultural activity, such as adverse climatological situations, pest or disease infestations, and marketing risks; in some cases, these are not fully covered by agricultural insurance or other risk management tools. To address this limitation, the Mexican Government has implemented the Program of Induction and Development of Rural Financing (Programa de Inducción y Desarrollo de Financiamiento al Medio Rural) to broaden the availability of financial services in rural areas. This program is directed toward financial intermediaries as well as people who traditionally have had difficulties obtaining loans from commercial banks. Federal resources for this program are dispersed in the form of Government guarantees, with maximum supports of 10 percent on total contracted credit, and up to 20 percent in highly marginalized and very highly marginalized regions. The program plays a fundamental role in fostering credit access for producer organizations that wish to sell crops housed in storage centers.

In a related effort, the Mexican Government is organizing dry bean producers using the framework of production systems (*sistemas producto*). Production systems are mechanisms of coordination among Mexico's Federal, State, and municipal governments that encourage the association and socioeconomic organization of producers and other economic agents involved in specific production chains. In the Production System for Beans, the development of activities is guided by a master plan that specifies short-, medium-, and longterm strategies; these activities range from the creation of statistical databases to competitiveness-enhancing activities, such as the consolidated purchase of inputs and advertising campaigns that promote dry bean consumption.[17]

Consistent with the objectives of the Production System for Beans, some dry bean producers receive support from the Program for the Sustainable Use of Natural Resources in Primary Production (Programa de Uso Sustentable de Recursos Naturales para la Producción Primaria). This program was first applied to the dry bean sector in 2004, after Mexico's National Institute for Forestry, Agricultural, and Fishing Research (INIFAP—Instituto Nacional de Investigaciones Forestales, Agrícolas, y Pecuarias) discovered that of the 1.2 million hectares planted with the crop in the States of Durango, San Luis Potosí, and Zacatecas during 1990-2003, 700 thousand hectares were located in areas of low or very low levels of potential productivity. This finding led to the implementation of a regional project that focused on the dry bean sector of these three States.

The Program for the Sustainable Use of Natural Resources in Primary Production features two main strategies for the dry bean sector. The first is to foster the use of improved seeds to produce varieties that are in greater demand, increasing domestic production and participation in the national market. The second strategy—sometimes referred to as "conversion" or "reconversion"—is to encourage dry bean farmers in areas with low or medium levels of potential productivity to switch to other crops that are in greater demand and have lower environmental impacts, such as a better use of water. This strategy converted about 477,000 hectares from 2004 to 2008. In place of dry beans, these areas now produce almost 980,000 mt per year of fodders, such as oats, triticale, and corn.[18]

Supporting investments in various aspects of the productive chain is another approach used by the Mexican Government to assist the dry bean sector. In 2007, the Strategic Project of Support for the Productive Chain of Corn and Bean Producers (PROMAF—Proyecto Estratégico de Apoyo a la Cadena Productiva de Maíz y Frijol) was established to support landholders in areas with medium and high productivity potential. To access the program, beneficiaries must: 1) prove that they are the legitimate owners of the land; 2) organize themselves into a producer's organization; and 3) contribute to the establishment of guarantee funds.[19]

PROMAF supports investment projects in corn and dry beans through two methods. The first method is to provide direct partial payments to finance technical assistance (70 percent of the project's cost), the strengthening of organizations (up to 60,000 pesos or about U.S. $4,800 per organization), and the support of newly graduated professionals (5,000 to 7,500 pesos or about U.S. $400 to $600 per month, depending on the area's poverty rate). This method also supports the acquisition of basic farm machinery and infrastructure by providing liquidity guarantees to producers in rainfed and irrigated areas.

Projects in zones with high or very high poverty rates receive guarantees up to 70 percent of the project's cost, with a limit of 2 million pesos (about U.S. $160,000) per project. In areas with low or medium poverty rates, each project receives a guarantee of up to 50 percent of its cost, also with a limit of 2 million pesos. The main types of acquisitions supported by this program include machinery for gathering, harvest, and post-harvest activities and irrigation equipment.

The second method is to support the application of technological packages in zones of medium or high productive potential using a risk-sharing approach. For first-time participants, PROMAF provides 25 percent of the package's cost, with a limit of 800 pesos (about U.S. $64) per hectare.

For returning participants, PROMAF provides 10 percent of the package's cost, with a limit of 400 pesos (about U.S. $32) per hectare, conditional on the Government's recovery of at least 60 percent of the support provided during the first year of participation. For all participants, the maximum amount of land that may be supported is 20 hectares of rain fed land and 5 hectares of irrigated land.

Table 5. Indicators of dry bean production by PROMAF participants during spring-summer 2008 and fall-winter 2008/09 agricultural cycles

	All dry bean producers			PROMAF participants only											
	Yields			Yields			Cost of production			Sale price			Cost of production		
Production cycle/State	2007	2008	Change	2007	2008	Change	2007	2008	Change	2007	2008	Change	2007	2008	Change
	Tons per hectare		Percent	Tons per hectare		Percent	Pesos per hectare		Percent	Pesos per hectare		Percent	Pesos per hectare		Percent
Spring-summer															
Total, Seven States	0.53	0.62	17	0.48	0.59	23	3,013	3,895	29	6,013	8,574	43	10,280	9,817	-5
Chihuahua	0.77	0.86	12	0.53	0.87	64	2,409	4,175	73	6,225	9,400	51	4,688	5,051	8
Coahuila	0.37	0.31	-16	0.42	0.39	-7	3,772	3,756	0	8,000	10,000	25	8,981	9,631	7
Durango	0.54	0.57	6	0.64	0.70	9	3,448	4,474	30	6,829	8,908	30	6,301	8,330	32
Jalisco	0.92	1.00	9	0.48	0.59	23	2,737	3,056	12	6,000	10,275	71	6,296	5,587	-11
Queretaro	0.52	0.90	73	0.55	0.53	-4	2,621	3,859	47	8,000	8,169	2	5,273	11,364	116
San Luis															
Potosi	0.36	0.65	81	0.31	0.52	68	3,068	3,927	28	5,388	8,216	52	16,871	12,751	-24
Zacatecas	0.52	0.57	10	0.45	0.54	20	2,562	3,298	29	5,898	8,618	46	8,485	9,653	14
Fall-winter															
Chiapas	0.59	0.61	3	0.68	0.90	32	4,008	4,534	13	8,557	10,922	28	6,315	6,086	-4

Source: State-level production data from SAGARPA/SIAP (2010a); PROMAF data from tables 18 and 19 in Colegio de Posgraduados (2009: p. 116, 118).

PROMAF= Proyecto Estratégico de Apoyo a la Cadena Productiva de Maíz y Frijol.

Notes: The Mexican production statistics in this report correspond to Mexico's agricultural years, which are divided into two production cycles: fall/winter and spring/summer. Yield data in this table are for the fall-winter 2007/08 production cycle, which was part of Mexico's 2008 agricultural year. All participants during the fall-winter 2008/09 cycle were located in the State of Chiapas.

Information on PROMAF's impact on the dry bean sector is gradually emerging. From 2007 to 2009, PROMAF provided improved technology and technical assistance to almost 1 million beneficiaries, corresponding to about 2.3 million hectares of corn and dry beans, 23 percent of which was cultivated with both crops (Secretaría de Agricultura, Ganadería, Desarrollo Rural, Pesca, y Alimentación, Fideicomiso de Riesgo Compartido, (SAGARPA/FIRCO) 2010: p. 8). On average, dry bean producers participating in PROMAF in 2008 obtained higher yields and lower costs of production on a per ton basis than in the previous year (table 5). These results differed substantially by State, however, with only participants in Jalisco and San Luis Potosí (during the spring-summer production cycle) and in Chiapas (during the fall-winter cycle) experiencing both an increase in their average yields and a decrease in their average costs (again on a per ton basis). On average, both participants and nonparticipants in Jalisco, San Luis Potosí, and Chiapas experienced yield increases between 2007 and 2008. Caution should be exercised when interpreting the program statistics in table 5, which reflect no effort to disentangle the influence of factors, such as favorable weather, that would affect PROMAF participants and nonparticipants alike or to address the representativeness of the participants.

U.S. Agricultural Income Supports Exclude Dry Beans

Dry beans do not belong to the set of "program crops" supported by the main agricultural income support programs—direct payments, marketing loans, and countercyclical payments—operated by the U.S. Federal Government. Although some pulses[20] were incorporated into the set of program crops starting with the 2002 Farm Act (Farm Security and Rural Investment Act of 2002) and continuing with the 2008 Farm Act (Food, Conservation, and Energy Act of 2008), dry beans were not among these crops.

Under the 2008 Farm Act, countercyclical payments, marketing assistance loans, and loan deficiency payments are available for dry peas, lentils, and chickpeas, while direct payments are not available for any pulse crops.[21] In addition, since some dry bean farmers have base acreage[22] for program crops other than pulses, they can participate in the income support programs associated with those crops.

Planting-flexibility restrictions within U.S. agricultural income support programs may actually limit the size of the U.S. dry bean sector. The 2008 Farm Act continues restrictions on the planting of fruits, vegetables, and wild rice (excluding mung beans, dry peas, lentils, and chickpeas) on base acreage. Dry beans are considered a vegetable within this framework, and the planting of dry beans on base acreage is prohibited unless the producer has a history of dry bean production.

Some research indicates that eliminating planting-flexibility restrictions would foster increased dry bean production in the United States. In a study focused on Michigan's fruit and vegetable sectors,

Thornsbury, Martinez, and Schweikhardt (2007) found that the barriers to entry associated with converting from program crops to dry beans were low in comparison with the other fruits and vegetables studied—pickling cucumbers, processed tomatoes, fresh-market tomatoes, squash, and blueberries. Thus, the authors concluded that ending the restrictions could lead to the entry of new dry bean producers in Michigan, a State that is already home to a prominent dry bean sector.

In a farm- and market-level analysis, Johnson et al. (2006) suggest that the effect of eliminating the planting-flexibility restrictions on dry beans would vary across regions, since the variability of dry bean revenue differs across regions. While the authors' partial equilibrium model of the U.S. dry bean sector showed that eliminating the restrictions would increase U.S. dry bean area by 27 percent and lower the sector's revenue by 13 percent, the authors emphasize that the increase in dry bean area by some farm program participants would be partially offset by reductions in dry bean area by some nonparticipants.

The Federal Government also operates a number of programs that broadly benefit the vegetable sector, including producers of dry beans (Lucier et al., 2006).

Examples of such programs include *ad hoc* disaster payments (Noninsured Crop Disaster Assistance Program—NAP), crop insurance, marketing and promotional assistance, food aid purchases, export promotion, and cost-share and other types of assistance for implementing conservation measures. The Federal Crop Insurance Corporation (FCIC) provides insurance coverage for dry beans through a program called Actual Production History (APH), which is designed to protect against yield losses due to natural causes. Under this program, producers decide the amount of average yield as well as the percentage of the predicted crop price that they wish to insure. If a producer's harvested yield is less than the insured yield, they are paid an indemnity based on the difference.

Table 6 summarizes APH coverage statistics for dry beans. These statistics suggest that a large portion of U.S. dry bean producers buy crop insurance. The net area insured corresponds to roughly 85-95 percent of the area planted, and the number of policies sold exceeds the number of dry bean producers tallied by the agricultural census, suggesting that some producers purchase multiple policies for different parcels of land. Little research has been conducted on the effects of crop insurance on the U.S. dry bean sector, but in general, crop insurance is thought to have a small, positive effect on area planted (see, for instance, Young, Vandeveer, and Schnepf, 2001).

On occasion, USDA will purchase dry beans and other agricultural commodities for use in nutritional assistance programs. For instance, in May 2009, USDA announced that it would buy up to $25 million of dry beans for this purpose, following a request from Nebraska's dry bean industry. By the end of June 2009, USDA's Agricultural Marketing Service (AMS) had purchased about 11,800 mt (valued at $12.2 million) of Nebraska's Great Northern beans—an amount that corresponded to 16 percent of the Nation's 2008 Great Northern crop.

The Great Northern crop of 2008 was unusually abundant—about 72,000 mt, compared with 54,000 mt in 2007.

Table 6. FCIC APH coverage and summary for dry beans, 2004-09

	2004	2005	2006	2007	2008	2009
Number of policies earning premium	7,485	8,665	7,893	7,116	6,600	6,882
		Hectares (thousands)				
Net hectares insured	490	614	584	555	517	544
		Dollars (millions)				
Liability[1]	206	266	277	292	411	429
Producer premium	12	16	17	18	25	27
Subsidy[2]	17	23	25	27	36	39
Total premium	29	39	42	45	61	66
Indemnity[3]	45	35	37	24	25	60

Source: USDA, Risk Management Agency (RMA) (2010).

APH=Actual Production History program.

Note: Data as of October 18, 2010.

[1]Amount insurer would pay in the event of a total loss.

[2]Amount of total premium paid by the Federal Crop Insurance Corporation (FCIC).

[3]Amount paid out by insurer based on loss suffered by insured.

Dietary Changes Shape the Prospects for Demand

Long-term prospects for U.S. and Mexican dry bean demand will depend on future dietary patterns in each country. In Mexico, the key questions yet to be answered about the outlook for dry bean demand are:

1. To what extent will the country's traditional diet based largely on white corn and dry beans persist?
2. What role will dry beans play in the nontraditional diets that emerge over time?

Currently, annual per capita consumption of dry beans is about 11 kilograms in Mexico, compared with 3 kilograms in the United States.[23] Dry beans are consumed throughout Mexico in households of every income level and in communities large and small (table 7); and something akin to the traditional diet is still prevalent in rural communities (less than 2,500 inhabitants), where households that indicate they are consuming dry beans report annual dry bean consumption of about 21 kilograms per capita.

Mexico's lower-income households tend to consume more dry beans. For members of these households, beans provide an important part of their daily intake of proteins, carbohydrates, and other nutrients, as well as a means for coping with higher food prices and/or lower earnings via increased bean consumption. Among Mexican households that reported consuming dry beans, the poorest decile (tenth) of households devoted 15.3 percent of their expenditures on proteins in 2008 to dry beans and other vegetables, compared with 6.8 percent for all households (INEGI, 2008).

Nevertheless, the long-term tendency is indicative of lower dry bean demand in Mexico. Between 1992 and 2008, real per capita expenditures on dry beans decreased at a compound annual rate of 2.7 percent (fig. 11). This decline is observable in all income deciles and in both rural and urban areas. Factors behind this decline include rising incomes, greater affordability of chicken and other meats, and increased consumption of alternative low-priced foods with a high caloric content, such as instant soups and noodles (Dickerson, 2005). Increased consumption of processed dry beans, such as canned beans and ready-to-eat meals that include dry beans, is offsetting some portion of the decreased consumption of dry beans in grain form.

Table 7. Annual Mexican consumption of dry beans in grain or processed form by households that indicated that they consumed dry beans, by size of community and income decile, 2008

Income decile	> 100,000	15,000-99,999	2,500-14,999	< 2,500	National total
		Kilograms per capita			
I	18.8	20.4	21.4	26.7	24.4
II	17.0	18.4	17.2	21.0	19.2
III	16.0	16.2	19.0	20.1	18.1
IV	14.8	17.0	16.6	24.1	18.1
V	16.3	16.2	14.6	16.5	16.0
VI	14.9	13.9	14.7	18.6	15.4
VII	14.4	15.6	14.0	17.1	15.0
VIII	14.3	15.2	13.4	16.9	14.7
IX	14.5	13.1	13.8	16.1	14.4
X	13.7	14.7	19.1	19.1	14.8
All deciles	15.0	15.8	16.4	21.4	17.1

Source: Author calculations based on data from Encuesta Nacional de Ingresos y Gastos de los Hogares (ENIGH) 2008, Instituto Nacional de Estadística y Geografía (INEGI, 2008).

Note: Decile I refers to the poorest 10 percent of Mexican households in terms of income, while decile X refers to the wealthiest 10 percent.

Although the long-term tendency is one of lower demand, Mexican expenditures on dry beans tend to increase during periods of economic difficulties. Table 8 compares changes in Mexico's gross domestic product (GDP) with changes in dry bean expenditures.

The three periods in the table in which dry bean expenditures expanded include years when Mexico's real GDP per capita contracted: -6.16 percent in 1995, -0.03 percent in 2001, and -1.65 percent in the fourth quarter of 2008 (compared with fourth quarter 2007).

Overall, the correlation between GDP growth and expenditure growth is negative (correlation coefficient of -0.79), which suggests that dry beans are an inferior good.[24]

This negative correlation also is present across income deciles for a given year: in general, the higher the income decile, the lower the household expenditures on dry beans.

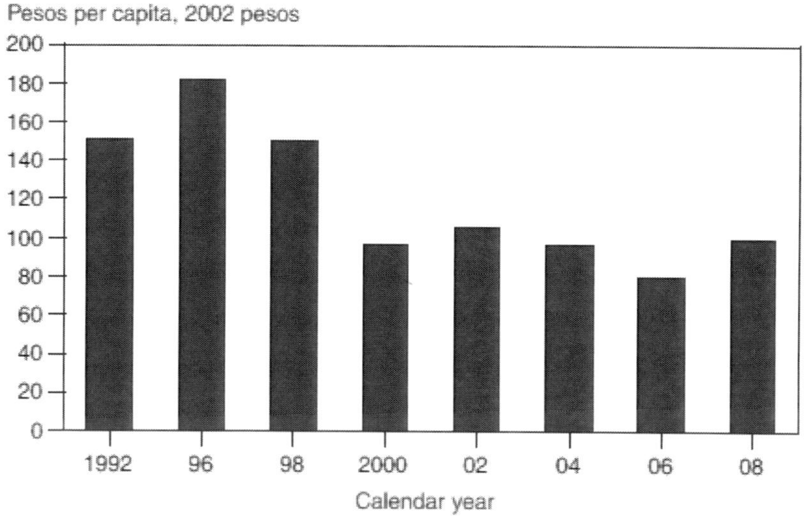

Sources: Author calculations based on expenditure data from Encuesta Nacional de Ingresos y Gastos de los Hogares (ENIGH) in Instituto Nacional de Estadística y Geografía (INEGI, 1992-2008) and population data from Consejo Nacional de Población (CONAPO, 2009).

Figure 11. Mexican household expenditures on dry beans, 1992-2008.

Table 8. Change in Mexican Gross Domestic Product (GDP) versus change in dry bean expenditures, 1992-2008

Period	Compound annual growth rate	
	Real GDP per capita	Dry bean expenditures per capita
		Percent
1992-96	1.24	4.68
1996-98	5.90	-8.98
1998-2000	5.16	-19.66
2000-02	0.37	4.42
2002-04	2.77	-4.24
2004-06	4.00	-9.18
2006-08	2.33	11.62
	Correlation	-0.79

Source: Author calculations based on real GDP data from Instituto Nacional de Estadística y Geografía (INEGI, 2009a), population data from Consejo Nacional de Poblacion (CONAPO, 2009), and expenditure data from Instituto Nacional de Estadística y Geografía (INEGI, 1992- 2008).

In the United States, the future of dry bean demand hinges largely on the extent to which people adopt diets that feature higher levels of dry bean consumption. U.S. dry bean demand is substantially lower than Mexican demand on a per capita basis. Annual per capita availability of dry beans in the United States peaked around 5 kilograms in 1942 and has never approached the level currently observed in Mexico (fig. 12). Still, U.S. dry bean producers face a challenge similar to that of their Mexican counterparts—little if any demand growth in their domestic market. During the early 2000s, annual per capita dry bean availability in the United States dropped as low as 2.7 kilograms, caused partly by the increased popularity of low-carbohydrate diets. At a national level, this change corresponded to a total decline in annual availability of roughly 150,000 mt (300 million people times 0.5 kilograms), or 13 percent of U.S. dry bean production in 2006.

U.S. dry bean demand has experienced a modest resurgence since 2004, with annual per capita dry bean availability stabilizing just under 3 kilograms. Factors behind this development include the declining popularity of low-carbohydrate diets, higher than average consumption by the country's growing Latino population, and efforts to promote dry bean consumption (Lucier and Jerardo, 2006). The most recent observation of annual per capita availability is 2.8 kilograms in 2008.

Further efforts to publicize the health benefits associated with dry bean consumption may hold the key to increasing U.S. dry bean demand.

The 2005 edition of the *Dietary Guidelines for Americans* recommends the consumption of 2-1/2 cups of vegetables per day for a reference daily intake of 2,000 calories, as well as the selection of vegetables "from all five vegetable subgroups (dark green, orange, legumes, starchy vegetables, and other vegetables) several times a week" (U.S. Department of Health and Human Services and U.S. Department of Agriculture, 2005: p. 24). Loss-adjusted data indicate that daily per capita vegetable consumption in the United States continues to be around 1.7 cups (USDA, Economic Research Service, 2010d).

Thus, increased per capita dry bean consumption would be one way to move toward the recommended level of vegetable consumption (Young and Cantor, 1999).

As a low-fat food, dry beans also can play a role in meeting dietary recommendations to "make choices that are lean, low-fat, or fat-free" when "selecting and preparing meat, poultry, dry beans, and milk or milk products" (U.S.

Department of Health and Human Services and U.S. Department of Agriculture, 2005: p. 30).

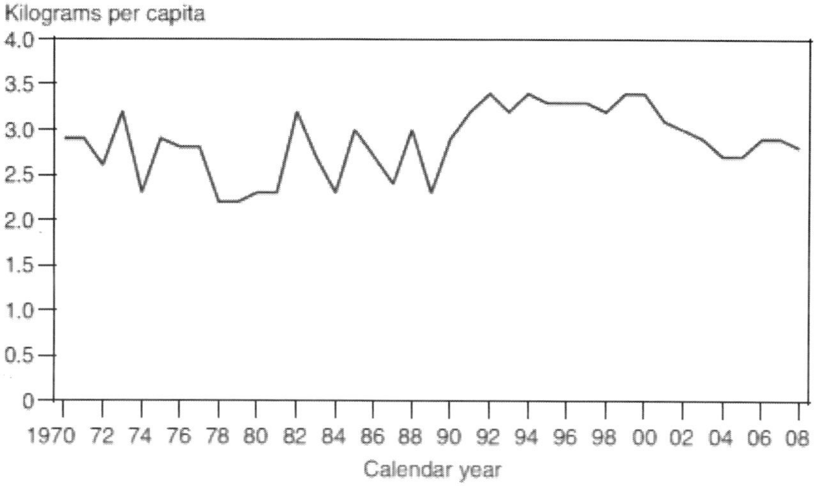

Source:Author calculations based on data from USDA, Economic Research Service (2010a).

Figure 12. Annual availability of dry beans in the United States, 1970-2008.

CONCLUSION

Liberalization of U.S.-Mexico dry bean trade concluded on January 1, 2008, with the removal of NAFTA's transitional TRQs. The gradual loosening of these trade restrictions during 1994-2007 enabled U.S. dry beans to provide a larger and steadier share of Mexican supply. Today, imports supply about 10 percent of Mexico's dry bean consumption, compared with 5 percent during the decade that preceded NAFTA (1984-93).

With the completion of NAFTA's transition to regional free trade, the outlook for the integrated North American dry bean market is shaped primarily by market forces and domestic agricultural programs, rather than anticipated reductions in tariffs and quotas. On the supply side, the restructuring of Mexico's dry bean sector will continue, most likely toward a smaller number of producers working larger plots of land. Recent improvements in yields, especially on irrigated lands, suggest that further yield growth is possible. The Mexican Government operates a number of programs designed to improve the competitiveness of the dry bean sector, focusing on direct income support, commercialization, access to credit, the adoption of superior technologies, and the organization of productive chains and production systems. Some of these programs have encouraged dry bean producers to pursue other agricultural activities instead. In the U.S. dry bean sector, the number of producers has steadily decreased over the past two decades, and the area devoted to the crop has stabilized over the last several years to around 534,000 hectares. Output growth is constrained by market opportunities as well as competition for land and other inputs. Recent yields exhibit a modest upward trend, due partly to favorable weather. U.S. producers who aspire to increase their dry bean sales outside the NAFTA region must contend with strong competition from other major producing countries, such as Burma (Myanmar) and China. Over the last several years, U.S. producers have had some success in recouping some of the previously lost share in the international navy bean market because of a contraction in Canadian production and a weaker U.S. dollar. Nevertheless, Mexico will remain a major destination for U.S. dry bean sales given Mexico's high levels of per capita dry bean consumption, the close proximity of Mexico to the United States, and the duty-free access to the Mexican market afforded by NAFTA. On the demand side, the prospects for dry bean producers and marketers are challenging in both Mexico and the United States. Per capita net disappearance of dry beans in Mexico is likely to decline even further from its current level of about 11 kilograms per year, as consumers diversify their diets and shift away from the traditional staples of beans and tortillas. Further declines in per capita consumption will limit the opportunities in Mexico's dry bean market for both Mexican producers and U.S. exporters. For the time being, however, beans are an indispensible source of protein and other nutrients for many Mexicans, particularly in households of modest means. Expenditure data suggest that beans are an inferior good in Mexico: consumption declines when income increases, and consumption increases when income declines.

In the United States, per capita annual availability of dry beans has rebounded to about 3 kilograms per year—about a fourth of the level in Mexico. Greater U.S. consumption of dry beans on a per capita basis could potentially have tremendous impacts on dry bean demand. For example, a 1-kilogram increase in annual U.S. per capita consumption would translate into an additional 300,000 mt in annual demand, requiring over 150,000 hectares in additional dry bean area. Over the long term, the demand-side challenge facing U.S. and Mexican dry bean sectors is fundamentally the same: how to cultivate demand for a product that is such a small dietary component for many higher-income consumers.

Appendix Tables

Appendix table 1. Tariff lines considered as dry common beans in this report

HTS code	Description
0713322000	Small red (Adzuki) beans (Phaseolus or Vigna Angularis), except seed, dried, shelled
0713333000	Navy or pea beans, except seed, dried, shelled
0713335000	Kidney beans, except seed
0713335020	Dark red kidney beans, except seed, dried, shelled
0713335040	Light red kidney beans, except seed, dried, shelled
0713335050	Pink beans, except seed
0713335060	Kidney beans, NESOI, including white pea beans (Phaseolus vulgaris), except seed, dried, shelled
0713335070	Kidney beans, except seed, other
0713395010	Black beans (Vigna spp., Phaseolus spp.), except seed,
	dried, shelled
0713395020	Great Northern beans (Vigna spp., Phaseolus spp.),
	except seed, dried, shelled
0713395050	Pinto beans (Vigna spp., Phaseolus spp.), except seed, dried, shelled
0713395060	White beans (Vigna spp., Phaseolus spp.), NESOI, except seed, dried, shelled
0713395070	Cranberry beans, except seed
0713395080	Beans (Vigna spp., Phaseolus spp.), NESOI, except seed, dried, shelled
0713395090	Beans, except seed, other

Source: Author compilation of tariff lines in USDA, Foreign Agricultural Service (2010).
HTS = Harmonized tariff schedule.
NESOI = Not elsewhere specified or included.

Appendix table 2. NAFTA's transitional TRQ for Mexican dry common bean imports from the United States and Canada, 1994-2008

Year	Over-quota tariff	Minimum duty-free quota, originating from:		Imports from United States		Imports from Canada	
		United States	Canada	U.S. statistics	Mexican statistics	Mexican statistics	Canadian statistics
	Percent			Metric tons			
1994	133.4	50,000	1,500	34,310	--	--	--
1995	127.8	51,500	1,545	17,712	24,049	696	68
1996	122.3	53,045	1,591	105,592	119,974	4,877	361
1997	116.7	54,636	1,639	61,892	86,649	1,685	60
1998	111.2	56,275	1,688	169,922	189,980	6,336	2,791
1999	105.6	57,964	1,739	73,006	121,619	1,736	1,918
2000	93.9	59,703	1,791	58,855	84,708	1,791	2,257
2001	82.1	61,494	1,845	72,860	115,675	8,600	5,526
2002	70.4	63,339	1,900	68,816	99,463	4,189	3,204
2003	58.7	65,239	1,957	61,001	76,232	4,389	2,117

Appendix table 2. (Continued)

Year	Over-quota tariff	Minimum duty-free quota, originating from:		Imports from United States		Imports from Canada	
		United States	Canada	U.S. statistics	Mexican statistics	Mexican statistics	Canadian statistics
	Percent			Metric tons			
2004	46.9	67,196	2,016	39,193	58,292	1,946	288
2005	35.2	69,212	2,076	62,681	74,323	1,867	1,169
2006	23.5	71,288	2,139	108,067	122,936	6,149	7,102
2007	11.8	73,427	2,203	79,994	85,341	4,113	3,309
2008	Free	Free	Free	99,110	87,339	5,486	3,867

TRQ = Tariff rate quota.

-- = Data not available.

Source: North American Free Trade Agreement (NAFTA) tariff schedule (TRQ information); Statistics Canada, as cited by Global Trade Information

Services, Inc. (2010) (Canadian trade statistics); Mexico, Secretariat of Economy (Mexican trade statistics); U.S. Department of Commerce,

Census Bureau, Foreign Trade Statistics, as cited by USDA, Foreign Agricultural Service (2010) (U.S. trade statistics).

Appendix table 3. Leading world exporters of dry beans: 1991-93 and 2005-07

Rank	Country	Annual average exports 1991-93	2005-07	Change
		Metric tons (thousands)		Percent
1	China	455	779	71
2	Burma (Myanmar)	376	511	36
3	United States	357	312	-13
4	Canada	90	302	237
5	Argentina	187	236	26
6	United Kingdom	13	53	319
7	Kyrgyzstan	--	41	--
8	Thailand	105	37	-65
9	Australia	16	36	130
10	Nicaragua	2	32	1,356
11	Ethiopia	*	32	81,988
12	Egypt	2	32	1,290
13	Colombia	9	27	194
14	Peru	3	26	671
15	Bolivia	8	25	199
16	Indonesia	2	22	999
17	Belgium	4	20	405
18	Uganda	24	19	-21

Rank	Country	Annual average exports 1991-93	2005-07	Change
		Metric tons (thousands)		Percent
19	Netherlands	32	18	-43
20	Mexico	8	18	109
	World total	1,976	2,755	39

Source: Food and Agriculture Organization of the United Nations (2009b).

-- = Data not available.

*Exports averaged less than 500 metric tons during 1991-93.

Notes: Countries in bold designate NAFTA countries. Data exclude broad beans, horse beans, and cowpeas. The definition of dry beans used by the Food and Agriculture Organization of the United Nations (FAO) differs somewhat from that used by USDA, so the data above do not necessarily match other figures in this report.

Appendix table 4. Leading world importers of dry beans: 1991-93 and 2005-07

Rank	Country	Annual average imports 1991-93	2005-07	Change
		Metric tons (thousands)		*Percent*
1	India	53	470	791
2	*United States*	*17*	*157*	*824*
3	Cuba	--	135	--
4	United Kingdom	95	121	27
5	Japan	105	120	14
6	Italy	41	101	148
7	*Mexico*	*13*	*101*	*702*
8	Brazil	91	89	-2
9	Pakistan	24	77	225
10	South Africa	54	70	28
11	Venezuela	60	68	14
12	Spain	35	54	53
13	Korea,			
	Republic of	25	52	105
14	Algeria	31	51	63
15	France	59	48	-18
16	China	58	48	-17
17	Malaysia	23	42	84
18	Kenya	19	40	109
19	Portugal	21	39	90
20	Costa Rica	1	38	3,091
21	Angola	30	37	23
22	Netherlands	67	36	-45
23	Turkey	3	35	979
24	Indonesia	45	35	-22
25	*Canada*	*11*	*35*	*206*
	World total	1,626	2,766	70

Source: Food and Agriculture Organization of the United Nations (2009b).

-- = Data not available.

Notes: Countries in bold designate NAFTA countries. Data exclude broad beans, horse beans, and cowpeas. The definition of dry beans used by the Food and Agriculture Organization of the United Nations (FAO) differs somewhat from that used by USDA, so the data above do not necessarily match other figures in this report.

Appendix table 5. Leading world producers of dry beans: 1991-93 and 2006-08

		Annual average production		
Rank	Country	1991-93	2005-07	Change
		Metric tons (thousands)		*Percent*
1	India	3,558	3,710	4
2	Brazil	2,673	3,363	26
3	Myanmar	471	2,501	430
4	China	848	1,305	54
5	*Mexico*	*1,128*	*1,167*	*3*
6	*United States*	*1,183*	*1,137*	*-4*
7	Tanzania	223	487	118
8	Uganda	404	433	7
9	Kenya	420	409	-3
10	Argentina	206	329	59
11	Indonesia	761	322	-58
12	*Canada*	*126*	*305*	*142*
13	Korea, Demo- cratic People's Republic of	310	300	-3
14	Rwanda	174	281	62
15	Iran	176	223	26
16	Burundi	341	206	-39
17	Ethiopia	32	201	536
18	Cameroon	90	200	122
19	Nicaragua	71	176	147
20	Pakistan	92	173	89
	World total	16,168	20,321	26

Source: Food and Agriculture Organization of the United Nations (2009a).

Notes: Countries in bold designate NAFTA countries. Data exclude broad beans, horse beans, and cowpeas. The definition of dry beans used by the Food and Agriculture Organization of the United Nations (FAO) differs somewhat from that used by USDA, so the data above do not necessarily match other figures in this report.

Appendix table 6. Area harvested, production, and yields of U.S. dry beans, by commercial class: Annual averages, 2007-09

Commercial class	Area harvested	Production	Yield
	Hectares	*Metric tons*	*Metric tons*
	(thousands)	*(thousands)*	*per hectare*
Navy	86	177	2.05
Great Northern	24	57	2.39
Small white	1	2	2.68
Cranberry	3	5	1.72
Pink	12	25	2.10
Light red kidney	21	42	2.05

Commercial class	Area harvested	Production	Yield
	Hectares	Metric tons	Metric tons
	(thousands)	(thousands)	per hectare
Dark red kidney	18	38	2.06
Black	70	132	1.88
Pinto	261	498	1.91
Small red	14	31	2.18
Total	527	1,041	1.98

Source: USDA, National Agricultural Statistics Service, Crop Production (January 2010).

Note: Total represents only those classes considered as common beans in this report.

Appendix table 7. Payment rates, target prices, and loan rates according to the 2008 Farm Act for calendar years 2010-12

Commodity	Direct payment rates		Target prices for countercyclical payments		Commodity loan rates	
	English measure	*Metric equivalent (per metric ton)*	*English measure*	*Metric equivalent (per metric ton)*	*English measure*	*Metric equivalent (per metric ton)*
Wheat	$0.52/bu	$19.12	$4.17/bu	$153.31	$2.94/bu	$108.09
Corn	$0.28/bu	$11.02	$2.63/bu	$103.54	$1.95/bu	$76.77
Grain sorghum	$0.35/bu	$13.78	$2.63/bu	$103.54	$1.95/bu	$76.77
Barley	$0.24/bu	$11.01	$2.63/bu	$120.64	$1.95/bu	$89.45
Oats	$0.024/bu	$1.66	$1.79/bu	$123.45	$1.39/bu	$95.86
Upland cotton	$0.0667/lb	$147.05	$0.7125/lb	$1,570.80	$0.52/lb	$1,146.41
Long-grain rice	$2.35/cwt	$51.81	$10.50/cwt	$231.49	$6.50/cwt	$143.30
Medium-grain rice	$2.35/cwt	$51.81	$10.50/cwt	$231.49	$6.50/cwt	$143.30
Soybeans	$0.44/bu	$16.18	$6.00/bu	$220.59	$5.00/bu	$183.82
Peanuts	$36/ton	$39.68	$495/ton	$545.64	$355/ton	$391.32
Other oilseeds	$0.80/cwt	$17.64	$12.68/cwt	$279.55	$10.09/cwt	$222.45
Dry peas	None		$8.32/cwt	$183.43	$5.40/cwt	$119.05
Lentils	None		$12.81/cwt	$282.41	$11.28/cwt	$248.68
Small chickpeas	None		$10.36/cwt	$228.40	$7.43/cwt	$163.80
Large chickpeas	None		$12.81/cwt	$282.41	$11.28/cwt	$248.68

bu = bushel; lb = pound; cwt = hundredweight; and ton = short ton.

Note: Converted to metric equivalent using conversion factors from USDA, National Agricultural Statistics Service (2009).

Source: USDA, Economic Research Service (2009).

REFERENCES

Agriculture and Agri-food Canada, Specialty Crops Section. *Statistical Overview of the Canadian Pulse Industry: 2008*, June 24, 2009, http://www.ats-sea.agr.gc.ca/can/4753-eng.htm.

Ag Weekly Online. "Bean Commission awarded specialty crop grant," October 5, 2009, http://www.agweekly.com/articles/2009/10/05/commodities/crop/crop41.txt.

Banco de México. *Índices de Precios al Consumidor y UDIS: Índice de Precios al Consumidor: Por Objeto del Gasto y Actividad Económica: INPC Nacional: Mensual*, 2010, http://www.banxico.org.mx/politicamonetaria-e-inflacion/estadisticas/inflacion/indices-precios.html.

Colegio de Posgraduados. *Proyecto Estratégico de Apoyo a la Cadena Productiva de los Productores de Maíz y Frijol 2008: PROMAF II: Acompañamiento Técnico e Indicadores de Impacto: Informe*, 2009 (circa).

Consejo Nacional de Población (CONAPO). "República Mexicana: Indicadores Demográficos, 1990-2050," spreadsheet, 2009, http://www.conapo.gob.mx/00cifras/00indicadores/00.xls.

Dickerson, Marla. "Steeped in a New Tradition," *Los Angeles Times*, October 21, 2005, http://articles.latimes.com/2005/oct/21/business/ fi-ramennation 21.

Figueroa Sandoval, Benjamin. "Engorda de ovinos con productos regionales: raciones a base de nopal deshidratado y frijol extrusado," presentation, 2009.

Food and Agriculture Organization of the United Nations (FAO). *FAOSTAT: Production: Crops*, 2009a, http://faostat.fao.org/site/567/default. aspx# ancor.

Food and Agriculture Organization of the United Nations (FAO). *FAOSTAT: Trade: TradeSTAT: Crop and Livestock Production*, 2009b, http://faostat.fao.org/site/535/default.aspx#ancor.

Global Trade Information Services, Inc. *World Trade Atlas*, internet version 4.7 b, Columbia, South Carolina, 2010.

Guzmán, Abel. "El Cultivo de Frijol en Durango," presentation at the 2010 International-U.S. Dry Bean and Other Specialty Crop Congress, Cancún, Quintana Roo, Mexico, February 12, 2010.

Idaho State Department of Agriculture. *International Trade and Domestic Market Development*, 2006, http://www.agri.state.id.us/Categories/AboutISDA/ annual_marketing.php.

Instituto Nacional de Estadística y Geografía (INEGI). *Banco de Información Económica*, 2009a, http://dgcnesyp.inegi.org.mx/bdiesi/bdie.html.

Instituto Nacional de Estadística y Geografía (INEGI). "Cuadro 8: Unidades de Producción con Cultivos en el Ciclo Primavera-Verano 2007 Según Superficie Sembrada, Cosechada y Producción Obtenida por Cultivo y Entitad Federativa," *Estados Unidos Mexicanos: Censo Agropecuario 2007: VIII Censo Agrícola, Ganadero, y Forestal*, Aguascalientes, Mexico, 2009b, http://www.inegi.org.mx/sistemas/TabuladosBasicos/ Default.aspx?c=17177&s=est.

Instituto Nacional de Estadística y Geografía (INEGI). "Cuadro 9: Unidades de Producción con Cultivos en el Ciclo Otoño-Invierno 2006-2007 Según Superficie Sembrada, Cosechada y Producción Obtenida por Cultivo y Entitad Federativa," *Estados Unidos Mexicanos: Censo Agropecuario 2007: VIII Censo Agrícola, Ganadero, y Forestal*, Aguascalientes, Mexico, 2009c, http://www.inegi.org.mx/sistemas/TabuladosBasicos/ Default.aspx?c=17177&s=est.

Instituto Nacional de Estadística y Geografía (INEGI). *Encuesta Nacional de Ingresos y Egresos de Hogares*, 1992-2008, http://www.inegi.org.mx/inegi/default.aspx?s=est&c=10656.

Johnson, Demcey, Barry Krissoff, Edwin Young, Linwood Hoffman, Gary Lucier, and Vince Breneman. *Eliminating Fruit and Vegetable Planting Restrictions: How Would Markets Be Affected?* U.S. Department of Agriculture, Economic Research Service, Economic Research Report No. 30, November 2006, http://www.ers.usda.gov/publications/err30/.

Juárez, Benjamin, and Mark Ford. *Mexico Grain and Feed Annual*, U.S. Department of Agriculture, Foreign Agricultural Service, Global Agricultural Information Network (GAIN) Report No. MX0017, March 12, 2010, http://gain.fas.usda.gov/Recent%20GAIN%20Publications/Grain%20and%20Feed%20Annual_Mexico%20City_Mexico_3-12-2010.pdf.

Lucier, Gary, and Alberto Jerardo. *Vegetables and Melons Outlook*, U.S. Department of Agriculture, Economic Research Service, Electronic Outlook Report VGS-314, April 2006, http://www.ers.usda.gov/publications/vgs/2006/04apr/vgs314.pdf.

Lucier, Gary, Rachael Dettmann, and Michelle Da Pra. *Vegetables and Melons Outlook*, U.S. Department of Agriculture, Economic Research Service, Outlook Report VGS-333, June 25, 2009, http://usda.mannlib.cornell.edu/usda/current/VGS/VGS-06-25-2009.pdf.

Lucier, Gary, Susan Pollack, Mir Ali, and Agnes Perez. *Fruit and Vegetable Backgrounder*, U.S. Department of Agriculture, Economic Research

Service, Electronic Outlook Report VGS-313-01, April 2006, http://www.ers.usda.gov/publications/VGS/apr06/vgs31301/vgs31301.pdf.

Rodríguez Gómez, María Guadalupe. *La Comercialización en la Cadena Agroindustrial del Frijol: Los Casos de Zacatecas y Sinaloa*, Guadalajara, Jalisco: Universidad de Guadalajara, 2006.

Secretaría de Agricultura, Ganadería, Desarrollo Rural, Pesca, y Alimentación, Fideicomiso de Riesgo Compartido (SAGARPA/FIRCO). *PROMAF 2010*, January 12, 2010, http://www.firco.gob.mx/promaf2010/ docs/Presen tacion_PROMAF_2010.pdf.

Secretaría de Agricultura, Ganaderia, Desarrollo Rural, Pesca, y Alimentación, Servicio de Información Agroalimentaria y Pesquera (SAGARPA/ SIAP). *Agricultura: Producción Annual: Cierre de la Producción Agrícola por Cultivo*, 2010a, http://www.siap.gob.mx/index. php?option=com_ wrapper&view=wrapper&Itemid=350.

Secretaría de Agricultura, Ganadería, Desarrollo Rural, Pesca, y Alimentación, Servicio de Información Agroalimentaria y Pesquera (SAGARPA/ SIAP). *Balanzas Disponibilidad-Consumo de Productos Agrícolas and Históricos de Balanzas Disponsibilidad-Consumo*, May 19, 2010b, http:// www.siap.gob.mx/index.php?option=com_content&view=article&id=58 &Itemid=379.

Secretaría de Economía. "Anuarios Estadísticos de Mercados Nacionales: Granos y Semillas," *Sistema Nacional de Información e Integración de Mercados*, 2010, http://www.economia-sniim.gob.mx/Nuevo/.

Stewart-Williams, Kathy. "Idaho Certified Bean Seed for Mexico," presentation at Expo Agro Chihuahua, August 14, 2009, http://www.expoagrochihuahua.com/expo/kathy%20stewart.pdf.

Thornsbury, Suzanne, Lourdes Martinez, and David Schweikhardt. *Michigan: A State at the Intersection of the Debate Over Full Planting Flexibility*, U.S. Department of Agriculture, Economic Research Service, Contractor and Cooperator Report No. 29, February 2007, http://ddr.nal.usda.gov/ dspace/bitstream/10113/33691/1/CAT31036247.pdf.

U.S. Department of Agriculture, Agricultural Marketing Service, Livestock and Seed Division, Market News Branch. *Bean Market News*, GL_ GR851, Greeley, Colorado, various issues, 2005-09, http://search.ams. usda.gov/mnsearch/mnsearch.aspx.

U.S. Department of Agriculture, Economic Research Service. *2008 Farm Bill Side-By-Side*, April 15, 2009, http://www.ers.usda.gov/Data/ Food Consumption/.

U.S. Department of Agriculture, Economic Research Service. *Food Availability: Spreadsheets: Legumes*, February 1, 2010a, http://www.ers.usda.gov/Data/FoodConsumption/FoodAvailSpreadsheets.htm#vegfr.

U.S. Department of Agriculture, Economic Research Service. *Nominal Monthly Country Exchange Rates*, September 22, 2010b, http://www.ers.usda.gov/Data/ExchangeRates/.

U.S. Department of Agriculture, Economic Research Service. *Real Monthly Country Exchange Rates*, September 22, 2010c, http://www.ers.usda.gov/data/exchangerates/.

U.S. Department of Agriculture, Economic Research Service. *U.S. Per Capita Loss-Adjusted Food Availability: Custom Reports*, March 19, 2010d, http://www.ers.usda.gov/Data/FoodConsumption/app/loss_adjusted.aspx.

U.S. Department of Agriculture, Economic Research Service, Farm Policy Team. *Farm and Commodity Policy: Glossary*, September 21, 2009, http:// www.ers.usda.gov/Briefing/FarmPolicy/Glossary.htm.

U.S. Department of Agriculture, Foreign Agricultural Service. *Global Agricultural Trade System*, 2010, http://www.fas.usda.gov/gats/Default.aspx.

U.S. Department of Agriculture, National Agricultural Statistics Service. *Crop Production*, various issues, January 1991-2010, http://usda.mannlib.cornell.edu/MannUsda/viewDocumentInfo.do?documentID=1047.

U.S. Department of Agriculture, National Agricultural Statistics Service. "Table 33. Specified Crops by Acres Harvested: 2007 and 2002," *2007 Census of Agriculture: United States Summary and State Data: Volume 1, Geographic Area Series: Part 51*, AC-07-A-51, December 2009a (updated version), http://www.agcensus.usda.gov/Publications/2007/Full_Report/Volume_1,_Chapter_1_US/st99_1_033_033.pdf.

U.S. Department of Agriculture, National Agricultural Statistics Service. *Crop Values: 2009 Summary*, February 2010a, http://usda.mannlib.cornell.edu/usda/current/CropValuSu/CropValuSu-02-19-2010_new_format.pdf.

U.S. Department of Agriculture, National Agricultural Statistics Service. *2007 Census of Agriculture: Quick Stats Ad-hoc Query Tool*, 2010b, http://quickstats.nass.usda.gov/.

U.S. Department of Agriculture, National Agricultural Statistics Service. "Table 34. Specified Crops by Acres Harvested: 2002 and 1997," *2002 Census of Agriculture: United States Summary and State Data: Volume 1, Geographic Area Series: Part 51*, AC-02-A-51, June 2004, http://www.agcensus.usda.gov/Publications/2002/Volume_1,_Chapter_1_US/st99_1_034_034.pdf.

U.S. Department of Agriculture, National Agricultural Statistics Service. "Table 42. Specified Crops by Acres Harvested: 1997 and 1992," *1997 Census of Agriculture: United States Summary and State Data: Volume 1, Geographic Area Series: Part 51*, AC97-A-51, March 1999, http://www.agcensus.usda.gov/Publications/1997/Vol_1_Chapter_1_U._S._National_Level_Data/us-51/us1_42.pdf.

U.S. Department of Agriculture, National Agricultural Statistics Service. "Table 42. Specified Crops by Acres Harvested: 1992 and 1987," *1992 Census of Agriculture: Volume 1, Chapter 1: U.S. National Level Data*, October 1994, http://www.agcensus.usda.gov/Publications/1992/v1-tbl42.pdf.

U.S. Department of Agriculture, Risk Management Agency (RMA). *Summary of Business Reports and Data: Pregenerated SOB Reports: National Summary by Crop*, October 18, 2010, http://www.rma.usda.gov/data/sob.html.

U.S. Department of Health and Human Services, and U.S. Department of Agriculture. *Dietary Guidelines for Americans 2005*, 2005, http://www.cnpp.usda.gov/Publications/DietaryGuidelines/2005/2005DGPolicyDocument.pdf.

U.S. Department of Labor, Bureau of Labor Statistics. *Consumer Price Index: CPI Databases: All Urban Consumers (Current Series)*, 2009, http://www.bls.gov/cpi/#data.

Vallejo Díaz, Jesús. "Producción de Frijol en Zacatecas," presentation at the 2010 International-U.S. Dry Bean and Other Specialty Crop Congress, Cancún, Quintana Roo, Mexico, February 12, 2010.

Vollrath, Thomas. *North American Agricultural Market Integration and Its Impact on the Food and Fiber System*, U.S. Department of Agriculture, Economic Research Service, Agriculture Information Bulletin No. 784, September 2003, http://www.ers.usda.gov/publications/aib784/.

Wilkins, Dave. "Trade Trip Provides Optimism on Seed Beans," *Capital Press*, September 6, 2009, http://cps.live.mediaspanonline.com/idaho/ dw-seed-beans-090409.

Wingett, Yvonne. "Latino Importer Knows His Beans," *Arizona Republic*, January 28, 2006, http://www.azcentral.com/arizonarepublic/news/ articles/0128beanimporters.html.

World Bank, Agriculture and Rural Development Team, Sustainable Agriculture Department, Latin America and the Caribbean Region, and Instituto Mexicano para la Competividad. *Mexico: Integration of the North American Market for Sensitive Agricultural Commodities: Policy Notes*

Overview and Dry Bean Market–Implications for Mexican Producers and Consumers, December 2007, http://siteresources.worldbank.org/ EXTLAC REGTOPRURDEV/Resources/503766-1225476272295/publicacion4.pdf.

Young, C. Edwin, and Linda Scott Cantor. *Moving Toward the Food Guide Pyramid: Implications for U.S. Agriculture*, U.S. Department of Agriculture, Economic Research Service, Agricultural Economic Report No. 779, May 1999, http://www.ers.usda.gov/publications/aer779/.

Young, C. Edwin, Monte L. Vandeveer, and Randall D. Schnepf. "Production and Price Impacts of U.S. Crop Insurance Programs," *American Journal of Agricultural Economics* Vol. 83, No. 5, Proceedings Issue (December 2001), pp. 1196-1203.

End Notes

[1] A TRQ is a quota for a volume of imports at favorable tariff rate. After the quantitative limit is reached, a higher tariff is applied to additional imports. In many TRQs, the in-quota tariff rate equals zero, and the over-quota tariff rate is prohibitively high.

[2] Matching tariff lines to dry bean varieties is not a precise science. For instance, the tariff line for small red beans (0713.32.2000) includes varieties that belong to the species Phaseolus vulgaris as well as Adzuki beans, which belong to the species Vigna angularis. Moreover, it is not certain that the tariff classifications have been consistently and correctly applied to the different varieties of dry beans over the years.

[3] Interest in the market opportunities for yellow beans in Mexico and the United States has motivated the Idaho Bean Commission to partner with researchers at Idaho State University and Oregon State University to develop virus-resistant yellow beans. This research was funded by a Specialty Crop Block Grant awarded to the Idaho Bean Commission by the Idaho Department of Agriculture and USDA (Ag Weekly Online, 2009). Some producers in Michigan are also growing mayocoba beans.

[4] Mexico's 2007 Agricultural Census (INEGI, 2009b, 2009c) suggests that the country's dry bean production is roughly 15 percent larger than the production statistics presented in table 1. According to that census, Mexico produced 1.1 million tons of dry beans during the 2007 agricultural year, compared with 994,000 tons as stated by the annual production statistics. If the census figures are more indicative of the sector's true size, then imports from the United States play a somewhat smaller role in the Mexican dry bean market. For Mexico's 2010/11 marketing year (January to December 2010), USDA's Foreign Agricultural Service (see Juárez and Ford, 2010) forecasts Mexican dry bean production of 1.1 million tons.

[5] The U.S. marketing year for dry beans runs from September to August. Thus, the 2008/09 marketing year ran from September 1, 2008, to August 31, 2009.

[6] Appendix table 2 lists the quotas and over-quota tariffs associated with NAFTA's TRQs for dry beans, along with the actual trade volumes governed by the TRQs. In-quota volumes under these TRQs received duty-free treatment.

[7] As members of the World Trade Organization (WTO), each NAFTA country generally is required to apply its MFN tariffs to all other WTO members. Important exceptions to the

MFN obligation include preferential trade agreements, such as NAFTA, and special access for developing countries.

[8] CUSTA liberalized almost all aspects of U.S.-Canada agricultural trade over the 9-year period that ended on January 1, 1998. CUSTA was subsumed by NAFTA in 1994.

[9] Appendix tables 3 and 4 summarize changes in dry bean exports and imports, respectively, for the leading exporting and importing countries between 1991-93 and 2005-07, while appendix table 5 describes changes in dry bean production among the leading producing countries between 1991-93 and 2006-08.

[10] The U.S. price series data collected by USDA's Agricultural Marketing Service (AMS) for dry beans may be divided into two main groups: bids to growers and dealer prices. For each major commercial class, there are usually price series data that correspond to the States that are the leading producers and marketers of that class, such as North Dakota and Minnesota for pinto beans and Michigan for black beans.

[11] Mexico's large wholesale markets (centrales de abasto) are the marketing channel for roughly 40 percent of the country's dry bean production. Both wholesale and retail activities take place at these markets, as many Mexican consumers buy produce directly from distributors at these locations.

[12] The Mexican production statistics in this report correspond to Mexico's agricultural years, while the U.S. production statistics correspond to calendar years. Mexico's agricultural year is divided into two production cycles: fall/winter and spring/summer. For dry beans, planting in fall/winter begins in October and ends in March, while the harvest begins in January and ends as late as September. Planting in spring/ summer begins in April and ends in September, while the harvest begins in June and ends as late as March. Thus, Mexico's 2008 agricultural year covers dry beans planted from October 2007 to March 2008 (fall/winter 2007/08) or from April to September 2008 (spring/ summer 2008).

[13] Appendix table 5 lists the world's leading dry bean producers, using information from the FAOSTAT database of the Food and Agriculture Organization of the United Nations (FAO, 2009a). Note that FAOSTAT's definition of dry beans differs somewhat from the definition used in the production statistics of the U.S., Canadian, and Mexican governments.

[14] For an assessment of the economic value of irrigation to the Mexican dry bean sector, see pp. 45-7 of World Bank, Agriculture and Rural Development Team, Sustainable Agriculture Department, Latin America and the Caribbean Region, and Instituto Mexicano para la Competividad (2007).

[15] Since many dry bean producers also raise crops other than dry beans, production statistics indicate the average dry bean area per farm, rather than the average size of a farm that grows dry beans.

[16] Loss rate refers to the ratio between area harvested and area planted. When expressed as a percentage, it is calculated by subtracting the ratio of area harvested to area planted from 1 and then multiplying the result by 100.

[17] More information about the Production System for Beans is available at http://www.sisprofrijol.org.mx/.

[18] A third possible strategy is to find new uses for dry beans. In an activity prompted by discussions in 2007 and 2008, just prior to the end of NAFTA's TRQs for corn and dry beans, USDA and SAGARPA financed efforts to encourage some dry bean farmers in the State of Zacatecas to use their product, blended with prickly pear cactus, to feed sheep imported from the United States (Figueroa Sandoval, 2009). Also, the U.S. Dry Bean Council has promoted several nutritional studies of tortillas made partly from dry beans.

[19] A guarantee fund consists of money allocated by SAGARPA that serves as collateral so that a prívate bank can lend money to a rural producer who otherwise would be ineligible.

[20] In North American agriculture, the term "pulse crops" commonly refers to dry (mature) peas, lentils, and small and large chickpeas (garbanzo beans) that are used as food by humans or as feed for animals.

[21] Appendix table 7 lists the payment rates, target prices, and loan rates associated with the 2008 Farm Act.

[22] Base acreage is defined as "a farm's crop-specific acreage of wheat, feed grains, upland cotton, rice, oilseeds, pulse crops, or peanuts eligible to participate in commodity programs" (USDA, Economic Research Service, Farm Policy Team, 2009).

[23] The Mexican statistic was calculated using estimates of dry bean consumption for food purposes from SAGARPA/SIAP (2010b) and population estimates from CONAPO for the period 2006-10; the U.S. statistic was obtained from USDA/ERS (2010a).

[24] Economists define a good as inferior when consumption is negatively related to income—when income rises, the quantity consumed decreases, and when income falls, the quantity consumed increases.

In: Global Agriculture. Volume 1 ISBN: 978-1-62257-369-1
Editors: M.R. Robertson, J.I. Schmidt

Chapter 3

INVESTMENT IN PROCESSING INDUSTRY TURNS CHINESE APPLES INTO JUICE EXPORTS*

Fred Gale, Sophia Huang and Yingying Gu

ABSTRACT

In this report, ERS profiles the remarkable growth in China's apple juice concentrate exports since the 1990s and analyzes the factors behind the boom. Investment in the juice processing industry brought apples from China to the world market in the form of apple juice concentrate. The United States is the largest buyer, and concentrate from China now accounts for two-thirds of the U.S. apple juice supply. China's juice industry sustained its dramatic growth by expanding into the country's hinterland with support from officials eager to develop the poor northwestern region. In recent years, rising apple prices have begun to act as a brake on the industry's expansion, and the industry faces challenges in improving the quality of apples used for juice processing.

Keywords: China, apples, fruit, juice, agricultural exports, prices, investment

---

* This is an edited, reformatted and augmented version of the United States Department of Agriculture Publication, FTS-344-01, dated October 2010.

INTRODUCTION

China has emerged as a major exporter of a wide range of food products (Huang and Gale; Lohmar et al.; Carter and Gunning-Trant). While China's competitive advantage in agricultural exports is often ascribed to its labor abundance, exports of its labor-intensive agricultural commodities are limited by the cost of shipping perishable products long distances and sanitary or phytosanitary concerns in overseas markets. Consequently, China's most internationally competitive commodities reach the world market as processed food products. Thus, capital investment in processing facilities is instrumental in bringing many of China's labor-intensive agricultural products to the world market.

In this report, ERS analyzes the emergence of China's apple juice concentrate (AJC) industry, a prime example of the nexus between capital investment and labor-intensive agricultural exports. A large expansion of orchard area made China the world's leading apple producer but also led to depressed prices in the 1990s.

A combination of government, private, and foreign investment built a network of apple juice processing plants that turned apples into juice concentrate that could be exported, creating a new market opportunity for China's vast supply of apples. China is now the world's leading exporter of apple juice concentrate, and the United States is the largest buyer. Two-thirds of the U.S. apple juice supply now comes from China.

Since the 1990s, China's AJC exports have boomed, fueled by the country's vast supply of apples and investments in the industry. During the 2000s, the AJC industry sustained its dramatic growth by expanding into China's poor northwestern region in search of additional apples to use as raw materials. Chinese officials encouraged the expansion by "guiding" investment and recruiting apple growers.

China is likely to remain the world's leading supplier of AJC because no other country can match its raw material supply. Juice companies, however, still must compete with the fresh apple market for a limited supply of fruit. The industry's potential for further expansion may be constrained by its ability to acquire apples for processing.

Rising apple prices may also act as a brake on the industry's expansion. Other challenges include the need to continuously improve the quality of apples used for juice processing.

China's Juice Gains Market Share

China first emerged as a significant exporter of apple juice concentrate in the late 1990s when it accounted for about 20 percent of world trade in AJC. Major exporters at that time included Poland, Germany, and other European Union countries, and Southern Hemisphere countries that supplied products during the Northern Hemisphere's offseason.

China's AJC exports grew at an astonishing rate, from under $50 million in 1998/99 to a peak of $1.4 billion in 2007/08. In 1999/2000, the value of exports more than doubled from the previous year and grew 20-40 percent annually until 2005/06. China's AJC export value then surged by $400 million in 2006/07 and by over $500 million in 2007/08. Exports fell by over 40 percent in 2008/09 due to declining orders and falling prices as a result of the global economic slowdown. China's AJC exports totaled $770 million in 2008/09, down 40 percent from the previous year but still 60 percent higher than in 2005/06.

China's emergence on the market represented a fourfold increase in world AJC trade. Sales by traditional exporters in the Southern Hemisphere and Europe remained relatively stable as China's sales exploded (figure 1).

The United States is the largest market for China's AJC exports. U.S. apple juice imports peaked at over $600 million in calendar year 2008 and exceeded $300 million in 2009. While China exported apple juice to about 60 countries during 2007-09, over 40 percent went to the United States, about 20 percent went to the European Union, and 10 percent went to Japan (figure 2). Japan was China's main market for AJC exports in the 1990s, but the United States and Europe accounted for most of the growth after 1999.

An initial surge of Chinese AJC exports in the 1990s prompted the United States to levy antidumping duties against some Chinese apple juice suppliers in 1999, but the action did not prevent U.S. imports of Chinese juice products from growing dramatically (Carter and Gunning-Trant; Warner).[1] ERS estimates that imports from China accounted for about two-thirds of the U.S. apple juice supply by 2007/08 (figure 3).

U.S. apple growers view China's AJC exports as a threat. U.S. apple industry leaders cite the rapidly rising volume and low cost of Chinese apple juice concentrate as the main force putting downward pressure on prices of juice and apples (Warner; U.S. Apple Association). At least one U.S. Senator has blamed "cheap Chinese apple juice concentrate imports" for harmful effects on U.S. apple farmers (Schumer).

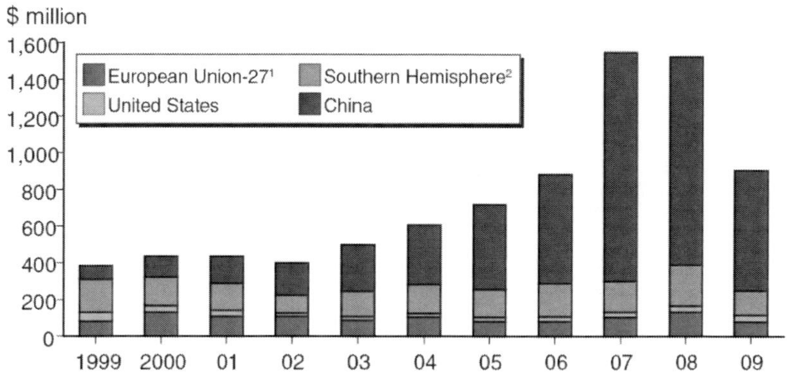

[1] External trade only; does not include trade between countries within the EU-27.

[2] Southern Hemisphere includes Argentina, Australia, Brazil, Chile, New Zealand, and South Africa. Data include Harmonized System codes 200970, 200971, and 200979.

Source: USDA, Economic Research Service analysis of customs data from Global Trade Information Services, Inc.

Figure 1. China is now the leading exporter of apple juice.

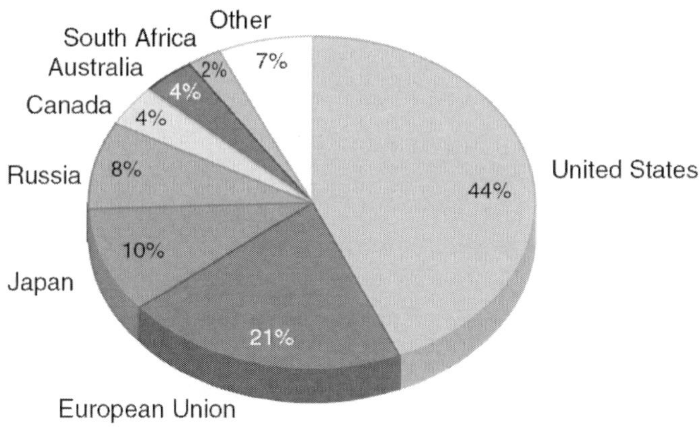

Note: Based on value of exports by destination during calendar years 2007-09 for Harmonized System codes 200971 and 200979.

Source: USDA, Economic Research Service analysis of customs data from Global Trade Information Services, Inc.

Figure 2. The United States is the largest market for China's apple juice exports.

The effects of increased AJC imports from China are more complex than a simple one-for-one displacement of U.S. juice. Manufacturers use juice and other ingredients with different degrees of acidity, flavor, and seasonal availability from China, the United States, and other countries. Chinese AJC may replace concentrate imported from other countries. It is possible that the availability of low-cost concentrate from China facilitated an expansion of the juice industry that would not have occurred otherwise, possibly expanding demand for U.S. juice apples if Chinese and U.S. juice concentrate are mixed in fixed proportions.[2] Competition from Chinese AJC also pushed U.S. growers to make innovations to cut costs and satisfy consumer demand (Hefling).

On balance, the rise in imports from China appears to have displaced roughly equal amounts of apple juice from domestic and other imported sources, but it also facilitated a net increase in apple juice consumption in the United States. From 2000/01 to 2008/09, U.S. apple juice imports from China rose by a cumulative 270,000 metric tons (mt) (see figure 3). During that period, the total U.S. apple juice supply from all sources (domestic production and imports) increased by 100,000 mt. The net expansion of the U.S. apple juice supply implies an increase in U.S. apple juice consumption because nearly all of the supply is consumed domestically.

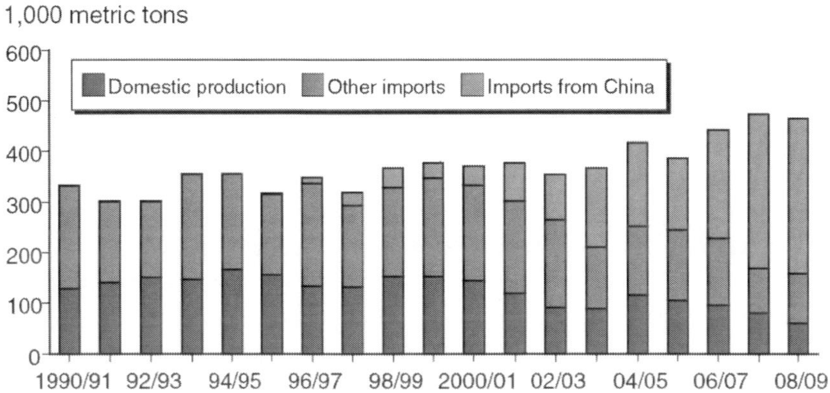

Note: Total imports and U.S. production from USDA supply and utilization estimates. Imports from China estimated based on China's share of import volume calculated from U.S. customs statistics. Market year is July-June.

Source: USDA, Economic Research Service estimates using data from USDA, Foreign Agricultural Service and Global Trade Information Services, Inc.

Figure 3. Imports from China now account for two-thirds of the U.S. apple juice supply.

Imports from China during the 2000s appear to have displaced a large volume of U.S. and other imported juice concentrate. The increase in imports from China (270,000 mt) far exceeded the cumulative increase in total supply (100,000 mt), which reflects a decline in apple juice supply from domestic sources of over 80,000 metric tons and a decline in imports from countries other than China (such as Chile, Argentina, Mexico, and the European Union) of about 90,000 metric tons.

BOOM IN CHINA'S APPLE PRODUCTION ENCOURAGED JUICE EXPORTS

Abundant, inexpensive fresh apples are the key to China's competitive advantage in producing apple juice. China began boosting apple production in the early 1990s as farmers sought to grow commodities that would bring higher returns than traditional field crops. In the early 1990s, China and the United States produced a similar volume of apples, each accounting for nearly a fifth of world apple production. After two decades of growth, China produced about seven times as many apples as the United States in 2009/10 and now accounts for roughly half of world apple production (figure 4).

Also during the 1990s, Chinese consumers diversified their diets to include more fruit, but not fast enough to absorb a fourfold increase in domestic apple production. Supply outpaced demand and pushed down Chinese apple prices. In 1991, the average farm price of apples in China was the equivalent of about $.25 per kg, roughly one-half of the U.S. price (figure 5). The price in China reached a low point of $.10 per kg in 2000, about one-fourth of the U.S. price that year. Chinese apple prices began rising after 2003, coinciding with the juice export boom. The average reached $.30 per kg in 2008—three times the price in 2000 but still about one-half the average U.S. price at the time.

Depressed apple prices in the 1990s encouraged agricultural officials in China to look to juice exports as an outlet for the apple supply. Low apple prices also generated interest among juice processors; according to Chinese industry sources, each kilogram of juice concentrate requires about 7 kg of apples as raw material and apples constitute 60-70 percent of AJC production costs (China Juice Concentrate Net).

In China, few apples are grown specifically for juice processing. The AJC industry offers an outlet for the abundant supply of apples that are too small, misshapen, or off-color to sell on the fresh market and would otherwise be discarded or used as animal feed. Typically, 20-30 percent of an apple crop is below fresh-market standards. USDA estimates indicate that processing accounted for only 5 percent of China's apple use in the 1990s, but the share rose to 15-25 percent in recent years.[3]

Apples used for juice processing command a much lower price than those destined for the fresh market. For example, according to an article from *Economic Reference News*, the highest quality apples for export in Shandong Province in 2009 brought a price of 3.4 yuan/kg (in U.S. dollars, 50 cents/ kg). Lower quality fruit for the domestic fresh market was quoted at about half that price, 1.6 to 1.8 yuan/kg (24-26 cents/kg). "Defective" or "fallen" fruit sold for juice processing brought much lower prices of .2 yuan or less (3 cents/kg or less). A farmer quoted in the article claimed that many traders come to the province to buy defective or fallen apples. "Even rotten fruit picked up off the ground can be sold," he said, "but the price is low" (*Economic Reference News*).

China's exports of both fresh apples and AJC have grown sharply since the 1990s (figure 6). The value of exports of fresh apples and AJC combined was roughly $100 million per year in the 1990s and rose to a peak value of $2 billion in 2007/08. The combined total was $1.47 billion in 2008/09. In most years, the export value from fresh apples and AJC were similar as exports of both products grew rapidly.

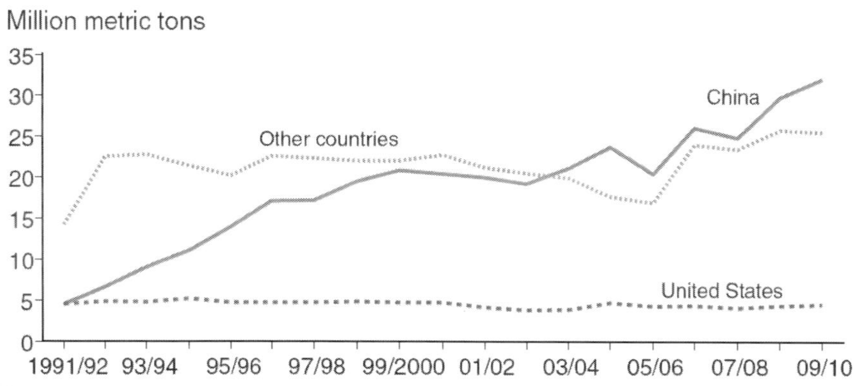

Note: Market year is July-June.

Source: USDA, Economic Research Service analysis of USDA, Foreign Agricultural Service data.

Figure 4. China now produces about one-half of the world's apples.

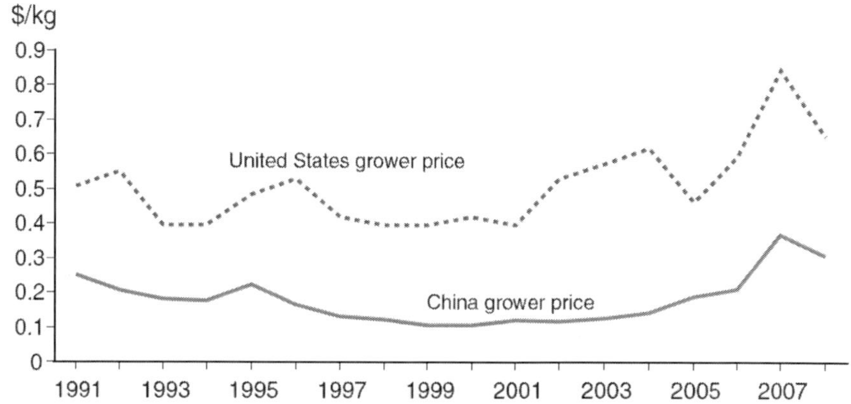

Note: Prices are season averages for fresh apples. Prices for juice apples are lower than those for fresh apples but follow similar trends. China prices are average prices received by farmers converted to dollars at the official exchange rate.

Source: USDA, Economic Research Service analysis of data from China National Development and Reform Commission and USDA, National Agricultural Statistics Service.

Figure 5. Low apple prices in China are key to juice competitiveness.

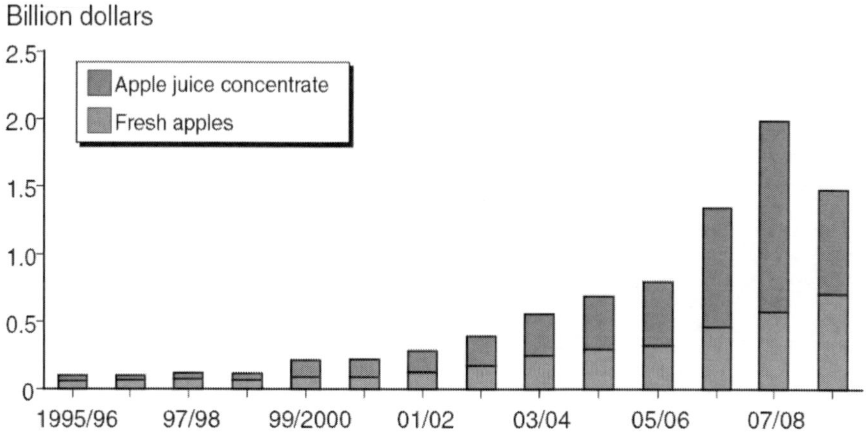

Note: Fresh apples include Harmonized System (HS) code 080810. Apple juice concentrate includes HS codes 200970, 200971, and 200979. Market year is July-June.

Source: USDA, Economic Research Service analysis of China customs statistics accessed through Global Trade Information Systems, Inc.

Figure 6. Combined value of fresh apple and juice exports reached $2 billion in 2007/08.

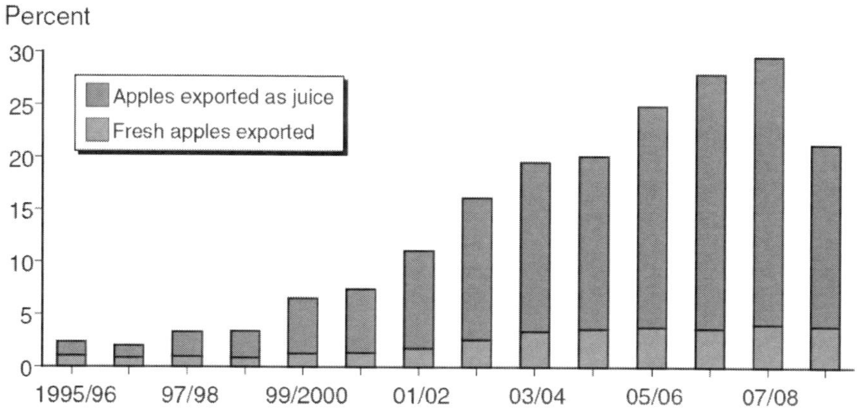

Note: Figure shows ratio of apples exported to apple production. Apples exported as juice estimated based on a ratio of 7 kg of apples required for 1 kg of apple juice concentrate. Market year is July-June.

Source: USDA, Economic Research Service estimates using data from China customs statistics and USDA Production, Supply, and Distribution database.

Figure 7. China's apple juice exports now consume a large share of its apple production.

AJC exports are now consuming a substantial share of China's apple crop. ERS derived estimates of the volume of apples required to produce AJC for export by multiplying the volume of AJC exports each year by the ratio of 7 kg apples required to produce 1 kg of AJC.[4] Based on these calculations, China exported the equivalent of 30 percent of its apple crop during the peak export year of 2007/08—26 percent for AJC and 4 percent for fresh apples (figure 7). The exported share fell to a combined 20 percent of the apple crop in 2008/09 when the global economic slowdown reduced international demand for juice. Still, this amount far exceeded the 2 percent of China's apple crop exported in 1995/96, which was split equally between exports of fresh fruit and apples used for exported juice.

Behind the Boom: Expansion of Processing Capacity

The transformation of apples into juice cannot take place without investments in processing plants and equipment. Thus, capital investment in processing capacity has been instrumental to the boom in China's apple juice exports. Apple growing is a labor-intensive process conducted by an estimated 10 million small farmers who have little education and minimal financial resources (China Ministry of Agriculture). Apple juice processing is a capital-intensive activity carried out by a few dozen companies that buy apples from farmers (through traders, brokers, cooperatives, or other intermediaries), crush them to make juice concentrate, and sell AJC mainly to large multinational companies as raw material for consumer products sold overseas. China's processing companies export over 90 percent of the AJC they manufacture.

Five companies account for over 70 percent of China's AJC exports (figure 8). The two leading exporters—SDIC Zhonglu Fruit Juice Co., Ltd. and Haisheng Group—each account for about 17 percent of exports. The other three leading companies have slightly smaller shares, and the remaining share is split among dozens of smaller companies.

Since the 1990s, the industry has expanded rapidly through a diverse mix of government, private, and foreign investment, loans from Chinese banks, and infusions of capital from stock market listings in China, Hong Kong, and overseas. The five major juice concentrate producers are conglomerates with diverse holdings that include regional subsidiary juice companies as well as businesses unrelated to apple juice.

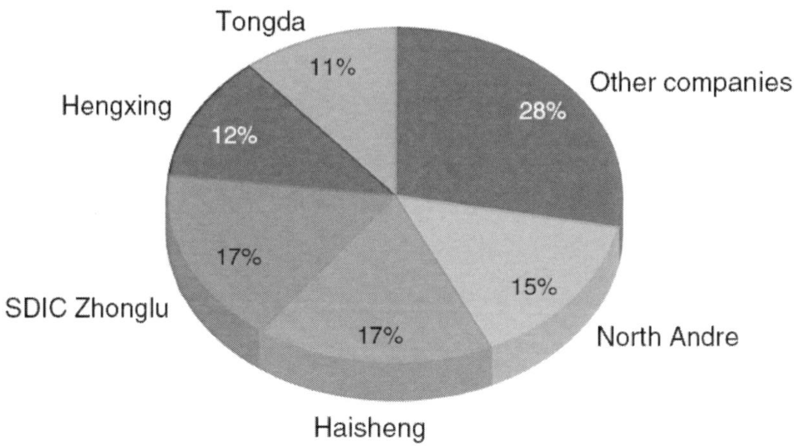

Source: USDA, Economic Research Service using Liu, Qian, *Review of 2008 Apple Juice Concentrate Industry and 2009 Outlook*, unpublished industry analysis report, March 23, 2009.

Figure 8. Five companies account for about 70 percent of China's apple juice exports.

China's apple juice industry was small and stagnant until 1992, when Japan's Mitsubishi Company formed a joint venture with a Chinese government investment company to set up a juice company in Shandong Province. The venture, SDIC Zhonglu Co., was China's first modern juice-processing company and began exporting AJC to Japan in the 1990s.

After the initial success of SDIC Zhonglu, a second wave of major companies were set up in 1996-97 and expanded rapidly. North Andre Fruit Juice Co., Ltd., founded in 1996, is also based in Shandong's main apple-producing area (Yantai City), but it now owns nine juice processing plants in five provinces. North Andre's initial owners were a diverse group of companies engaged in the fruit business, afforestation projects, and the supply of construction materials. Two other leading juice companies, Hengxin Fruit Juice Co., LTD. and Haisheng, were founded in 1996-97 in Shaanxi Province, another apple-producing region. Like Zhonglu, Hengxin was set up as a joint venture with a 51-percent controlling interest held by a Singapore company (QAF). Haisheng is a holding company registered in the Cayman Islands with shares owned by a diverse group that includes individual Chinese investors, another Chinese holding company, and the U.S. investment bank Goldman Sachs. The Haisheng group consists of local fruit juice subsidiaries, an investment company, and a U.S. sales branch.

Another wave of investment flowed into the industry in the first half of the 2000s. In 2002, Tongda Fruit Juice and Beverage Co. Ltd., the fifth-leading AJC company, was established. North Andre was listed on the Hong Kong Stock exchange in 2003, giving it access to more capital. In 2004, Zhonglu raised funds through an initial public offering in China. Most of the investment at this stage was domestic, but foreign investment also played a role. Goldman Sachs acquired a passive 20-percent ownership stake in Haisheng in 2005, and the International Finance Corporation (the investment arm of the World Bank) made investments in North Andre to help the company expand its operations.

The industry expanded rapidly as each company sought to maximize its share of a growing market. Companies built new production lines and acquired smaller companies. According to industry reports, China's apple juice concentrate production capacity rose from about 3,000 mt per year in the 1980s to 20,000 mt in 1995. As new companies entered and expanded, capacity shot up to 370,000 mt in 2001, and it expanded 44 percent in 2004/05 alone (Chen). Recent industry reports indicate that capacity now far exceeds export demand (China Juice Concentrate Net).

Until the late 2000s, China's juice industry was segmented between the export-oriented "upstream" AJC processors described in this section and a different set of "downstream" companies, including Huiyuan, Coca Cola, and President, that serve the domestic consumer beverage market. More recently, with heated competition and maturation of the domestic market, some AJC companies have sought to forward-integrate into beverage products. Consumer beverage companies have begun backward-integrating by forming joint ventures with processors or developing their own supply networks. The Huiyuan Co. has begun a strategy of contracting with "production bases" of farmers who can supply fruit as raw materials. Also in 2009, COFCO, a large state-owned food company, signed an agreement with authorities in Xinjiang Autonomous Region to invest 2 billion yuan ($290 million) over 5 years in fruit production and processing and marketing of juice and other fruit products.

Regional Shift to the Northwest

Geography is an important factor in the development of the juice industry. Apple juice production is located mainly in China's two main apple-producing regions: the Bohai Gulf region (Shandong, Liaoning, and Hebei Provinces) and the Loess Plateau region (Shaanxi, northwestern Henan, southwestern Shanxi, and parts of Gansu and Ningxia Provinces) (figure 9).[5] The government's strategic plans for the apple industry call for concentrating production in these two regions based on their temperature, rainfall, and soil type.

The Bohai Gulf region is on China's east coast, where easy access to Japan and South Korea encouraged the initial round of apple juice investment and trade in the 1990s. But further expansion in that region was constrained by scarcity of land and labor due to the rapid development in both agriculture and industry in eastern China. Recent growth in apple production is mostly in the northwestern provinces of the Loess Plateau region. Studies of climate and soils show that this region has excellent conditions for growing apples. Farmers and officials in the Loess Plateau are encouraged by the potential positive effects that apple production may have on the local economy, particularly as industrial development in the region has been slow and few other crops grow well due to the region's arid climate and hilly topography.

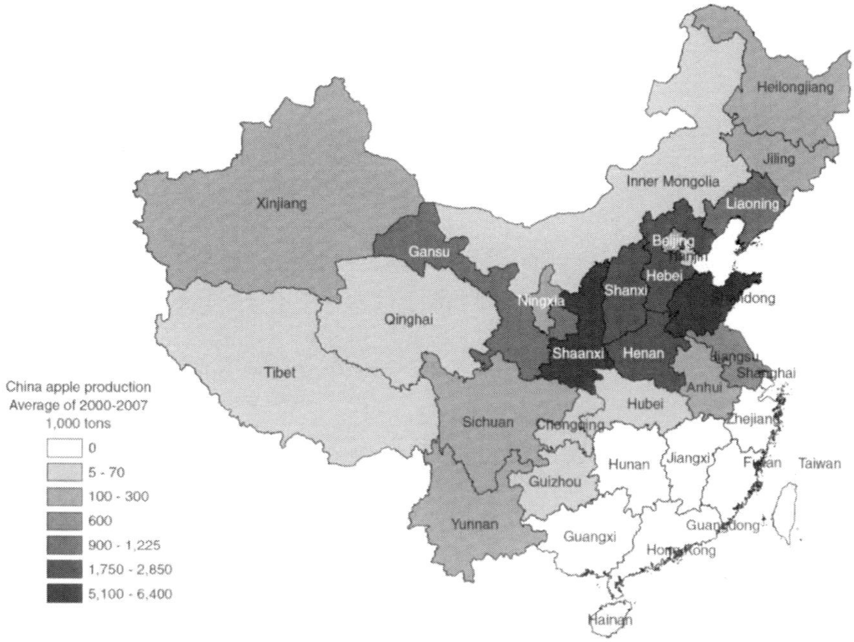

Source: USDA, Economic Research Service analysis of data from China National Bureau of Statistics.

Figure 9. China's leading apple-producing regions are Shandong and Shaanxi Provinces.

The Loess Plateau region accounted for most of the increase in China's apple production between 2000 and 2008. Shaanxi's production nearly doubled during 2000-2008, and its apple output is now comparable to that of Shandong, traditionally the leading apple-growing province.[6] Over the same period, production also increased substantially in Henan, Gansu, and Shanxi (figure 10).

Analysis of Chinese customs statistics by region shows that most of China's apple juice exports come from a belt of counties in the Loess Plateau and eastern Shandong and Liaoning Provinces, which reflects the geographic concentration of processing investment (figure 11). Tabulations of customs data for calendar year 2009 show that two-thirds of China's apple juice exports came from provinces in the Loess Plateau region. The leading exporter was Shaanxi Province (43 percent of exports). Other Loess Plateau provinces supplying juice to the international market included Henan (10 percent of exports, mostly from the western part of the province), Shanxi (7 percent), Gansu (6 percent), and Ningxia (1 percent). Provinces in the Bohai Gulf region—Shandong (25 percent of exports) and Liaoning (6 percent)—accounted for 31 percent of apple juice exports.

ERS estimates that Shaanxi's juice exports required over 2.3 million metric tons of apples as raw materials (table 1), equivalent to over one-fourth of the province's apple harvest in 2009. Shaanxi also was China's second-leading source of fresh apple exports, but the province's fresh apple exports were only about one-tenth of the quantity used to produce its juice exports. Henan and Shanxi exported few fresh apples. The volume of apples required for Gansu's juice exports was equivalent to nearly 10 times the volume of its fresh apple exports.

Most of China's fresh apple exports come from the Bohai region, but this region also uses more apples for juice exports than for fresh exports. Shandong is the leading source of fresh apple exports—Shandong exports twice as many apples as Shaanxi—but Shandong's apple use for juice exports is more than double the amount of fresh apples it exports.

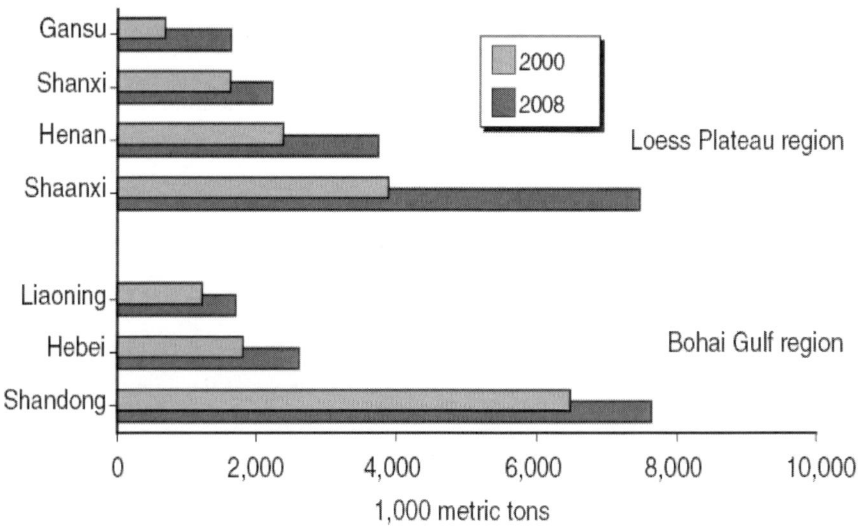

Source: USDA, Economic Research Service using data from China National Bureau of Statistics.

Figure 10. The Loess Plateau region accounted for most of the increase in China's apple production in 2000-2008.

Table 1. Estimated apples for juice and fresh exports, 2009

Province	Apples exported as:	
	Apple juice	Fresh apples
	1,000 metric tons	
Northwest:		
Shaanxi	2,377	221
Henan	542	<10
Shanxi	399	<10
Gansu	350	36
Bohai Gulf:		
Shandong	1,373	549
Liaoning	344	75
Hebei	<10	16
Other regions:		
Xinjiang	<10	90
Heilongjiang	0	83
Other provinces	186	93

Note: Based on customs statistics by source of exports. Apples for juice exports reflects estimate of fruit required as raw material to produce quantity of apple juice exported. Calculated assuming 7:1 ratio of apples to juice. Data are for calendar year.

Source: USDA, Economic Research Service estimates based on China customs statistics accessed through Global Trade Information Services.

GOVERNMENT SUPPORT TO "DEVELOP THE WEST"

Government support played a role in promoting the northwestern region's apple juice industry development. In 2000, the Chinese government formally launched a "Develop the West" (*Xibu Da Kaifa*) program to steer investment to the less-developed western provinces, including the Loess Plateau (Ma and Summers). The program includes direct government investment, fi nancial transfers to western provincial governments, and such measures as subsidized loans, tax breaks, and access to infrastructure projects to "guide" private investment to western provinces.

The strategy emphasizes developing industries based on local resources ("industries with special characteristics"). The Ministry of Agriculture issued a set of strategic plans ("advantaged regional layout plans") that identified apples as one of China's most internationally competitive crops and called for concentrating production in the most efficient production regions, including the northwestern provinces.[7] In 2006, the Ministry of Commerce formulated a plan for promoting agricultural exports that encouraged companies to develop "export bases" in inland areas.

According to news articles and documents, investment in the apple and apple juice industries was encouraged as part of the Develop the West initiative. Shaanxi Province cites the quadrupling of apple juice concentrate exports as one of the initiative's achievements. Other news articles point to North Andre's expansion into northwestern provinces as an expression of support for Develop the West (Fuquan Zhao). IFC describes its investment in North Andre's ventures in the northwest as motivated by poverty alleviation.

Shaanxi's third-largest apple juice company, Tongda Group, was explicitly set up as part of the Develop the West program.[8] Tongda was established with investment by Shenzhen Eastern Development Group, a real estate conglomerate based in Shenzhen, China's most wealthy city, which is engaged primarily in businesses unrelated to apple juice (construction of housing projects, hotels, shopping malls, and infrastructure). The Tongda Group website highlights the contributions it has made to improving the welfare of apple farmers in several remote counties of Shaanxi, Gansu, and Ningxia Provinces. The Tongda website also includes a paragraph describing the company's contract to construct a sewage treatment plant for a county government in Gansu Province, and the parent company's web site mentions that the company built a shopping mall in Xian, the capital of Shaanxi.

Provincial and local governments have supported the industry's growth in the northwest by working with local officials to recruit apple growers, disseminating new technologies and apple varieties, building breeding centers and infrastructure, assisting farmers and companies in adopting safety standards, arranging bank loans, giving tax breaks, awarding food safety certifications, and setting up market information networks. In 2000, Shaanxi province officials issued a document promoting the fruit industry—especially apples for processing—as an engine of rural economic development. Shaanxi was the first province to establish a Fruit Industry Bureau, and provincial officials formulated a plan to expand apple production more than 50 percent by 2012 (*Invest in Shaanxi*; Guangfei Zhao). Local branches of China's inspection and quarantine bureaus assist processing plants and farmers in attaining certifications and registrations needed to make sales in overseas markets (Sanmenxia Municipal Government).

Note: Based on tabulation of exported quantity of apple juice (Harmonized System codes 200979 and 200971) during 2009 reported by multicounty regions under the administration of cities.

Source: USDA, Economic Research Service using data from China customs statistics accessed through Global Trade Information Services.

Figure 11. China's **apple juice exports** originate in its Loess Plateau and Bohai Gulf regions.

Integrating juice companies with small farmers is part of the government's development strategy. Shaanxi's development plan was to "follow the large company development road" by cultivating "dragon head enterprises" that lead farmers into the market (China Ministry of Agriculture Development Plan Office, pp. 386-87). Dragon head enterprises (the term is sometimes translated as "leading" or "flagship" enterprises) are companies identified by central, provincial, or local governments as having potential to help farmers by providing them with a market for their products and disseminating technical information. In return, companies receive tax breaks, preferential bank loans, access to land, assistance raising capital, and government help meeting export standards. An official of the Shaanxi Fruit Industry Bureau identified dragon head enterprises as the key players in the fruit industry and listed support measures that included low-interest loans, subsidies for storage and cold-chain facilities, and export incentives such as value added tax rebates for exports (*Invest in Shaanxi*).

Government officials sometimes act as intermediaries by introducing overseas customers or arranging trade shows. Shaanxi's industry plan called for local juice companies to unite with farmers to sell in the international market. A similar plan for Henan Province called for government investment of 20 million yuan for building five model refrigerated apple storage facilities and 40 million yuan for building four model juice factories. China's Export-Import Bank earmarked special loans to help Shaanxi juice companies with cash flow because farmers want to be paid in cash (Tie).

PRICES ON THE RISE

After more than a decade of rapid growth in China's juice industry, competition to procure China's apple supply has intensified. The price of apples is rising, and concerns have been raised about the quality of the apples themselves. A trend toward higher prices that began in 2004 was punctuated by an unusual spike in juice and apple prices in late 2007. Recent industry reports indicate that the limited supply of juice apples prevents processors from fully utilizing capacity.

The average unit value of Chinese apple juice exports—a proxy for the price of juice exports—rose from a low point of about 60 cents per kilogram in 2003 to about $1 per kg in 2007 (figure 12). In late 2007, the unit value shot up as high as $2 per kg as a poor apple harvest in Europe tightened world juice supplies at a time when most commodity prices were rising. With a surge of orders for Chinese AJC, panic buying of apples pushed the price of 7 kg of juice apples (the amount needed for 1 kg of juice) to as high as $1.70 in late 2007.[9] By comparison, the wholesale market price of apples (which reflects domestic fresh market prices) did not display any unusual increase during this period, suggesting that the panic-buying phenomenon was confined to the juice-apple market.

The surge in juice prices ended abruptly in 2008 when China's apple harvest surpassed expectations, the world economy began to slow, export orders plummeted, and many juice processors idled their plants. Prices for exported juice and juice apples dropped in late 2008. Industry news reports recounted piles of unsold apples in villages and warehouses in 2008 (Li et al.;

Nongbo Net, 2009; Wang). The wholesale apple price also dipped in late 2008, another reflection of the slowdown in the general economy.

After the boom and bust had run its course, prices settled at a level similar to their pre-boom values. In the latter half of 2009, the unit value of juice exports was roughly $.80 per kg, lower than the $1-per-kg value preceding the boom in 2007 but about one-third higher than the average in 2002-03. The cost of juice apples dropped as low as $.20 per kg in 2009 but recovered to $.60 in early 2010, similar to the pre-boom level and well above the 2004 value. The wholesale price of apples also rose to about $.70 per kg in 2010, about double the average price in 2002-03.

Addressing Quality and Safety Issues

The supply of raw materials is also a major determinant of the quality and safety of juice concentrate. A weakness of Chinese juice in the international market is its low acidity, which reflects the predominance of sweet apple varieties like Fuji and Guoguang used as raw material. Sweet juice from China must be mixed with other juices and ingredients to make final products with high acidity demanded by consumer markets in North America and Europe.

A number of industry representatives in China have voiced concerns over the lack of high-acid apples. In Shaanxi Province, high-acid varieties account for less than 10 percent of apple area, and juice processors there procured less than half the volume of high-acid apples needed to fulfill their contracted juice sales in 2005 (*Economic Reference News*; Xuetao Liu). Lacking options, processors use sweet apple varieties intended for the fresh market that yield juice with low acidity; the resulting juice concentrate may sell at a 40-percent discount or be rejected by customers (*Economic Reference News*). An industry expert quoted in another article described the lack of high-acid apples as a bottleneck for the industry's growth (Nongbo Net, 2009).

While the price for apples purchased by juice companies has been rising, apples sold for the fresh market still sell for higher prices. Consequently, the expansion of high-acid apple plantings has been slow because farmers prefer to grow sweet varieties that can be sold on the fresh market. Some companies and officials are promoting planting of orchards of high-acid apple varieties specifically for use in juice processing (Nongbo Net, undated) and the government's strategic plan for the apple industry encouraged production of these varieties (China Ministry of Agriculture). However, farmers are reluctant to plant apples primarily for juice processing. Juice processors use predominantly defective or fallen fruit intended for the fresh market. While pasteurization kills bacteria, Chinese industry experts warn that use of poor quality fruit can yield juice with a brown color or poor taste, or juice with high levels of patulin (a mycotoxin produced by molds) that is rejected by buyers (*Economic Reference News*; Yue Cui).

Some U.S. consumers voice concerns over potential food safety risks of juice from China (Robinson-Jacobs), but to date there have been no food safety incidents involving Chinese juice (CCCFNA; Gale and Buzby) despite the low quality of fruit used as raw materials. Manufacturers and officials in China closely monitor overseas safety standards because over 90 percent of their products are exported. Processing plants use high-quality imported processing equipment and quality management systems required by overseas customers. Chinese universities have also been active in developing methods to detect and eliminate potentially harmful pesticide residues and toxins from apple juice and implementing traceability and recall systems (Pu et al.) Overseas industrial customers carry out third-party audits and conduct their own testing of juice products as well.

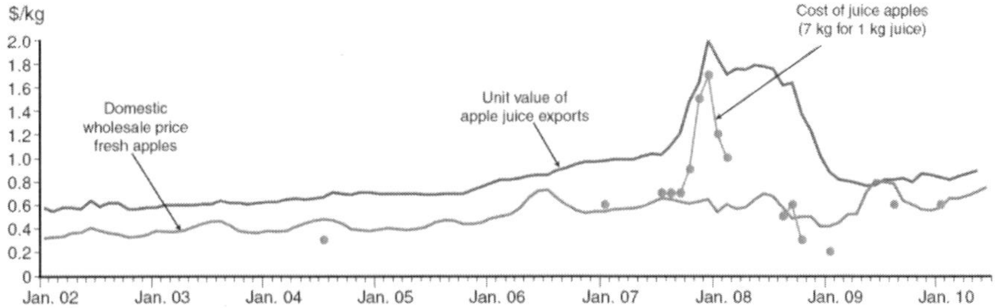

Note: Unit value of apple juice exports is ratio of value of exports to volume. Cost of juice apples is based on requirement of 7 kg of apples for 1 kg of juice; prices gathered from news reports. Domestic wholesale price of fresh apples is average price in farm production areas; reflects averages mainly for apples sold for fresh consumption. China data converted to U.S. dollars using official exchange rate.

Source: USDA, Economic Research Service analysis of data from China customs statistics, China National Bureau of Statistics, and industry news reports.

Figure 12. China prices of exported apple juice and apples display upward trend.

Aware of the risks and potential negative consequences of food safety incidents, Chinese officials have been on guard against food safety problems. In 2007, China's National Native Produce and Animal By-products Import and Export Association (an arm of the Foreign Trade Ministry) convened a meeting where representatives of major companies signed a pledge that they would adhere to good food safety practices throughout the supply chain (CCCFNA). The meeting followed the appearance of an article in the Chinese press alleging that one company processed rotten apples to manufacture juice. The document included a pledge to "...resist diseased and rotten fruit" (Yue Cui).

Concerns about chemical residues and adulteration of juice apples are reflected in an article reporting an inspection of apple-growing and juice-processing areas in Shaanxi Province (Zhou and Guan). The authors recommended that local officials train farmers with emphasis on the slogan "don't sell inferior [fruit]; don't adulterate." The authors also urged police to crack down on rogue apple merchants, and local governments were urged to send technical advisors to help farmers apply pesticide and fertilizer and carry out strict inspections of farm input dealers to "prevent fake pesticides and fertilizer from entering orchards."

China's inspection and quarantine authority (AQSIQ) and its provincial and local branches (known as CIQs) test and monitor exported products but are also engaged in helping companies and orchards meet overseas standards. Companies are expected to adopt measures to ensure their products meet overseas standards, but CIQs inspect facilities and test products prior to export. CIQs advise exporting companies about standards on foreign markets, inspect the layout of new or refurbished facilities, inspect orchards that supply fruit, and help companies implement quality management systems required by overseas customers (Henan MOFCOM; Sanmenxia Municipal Government; Shaanxi CIQ).

The articles about CIQ activities report that pesticide residues on China's apples are a longstanding concern for exporters of fresh-market apples and juice. In 2007, the CIQs in Henan and Shaanxi conducted pesticide surveys, required companies to implement remediation measures, and set testing procedures and thresholds for exports to various countries. The Henan articles report that no juice export shipments had been rejected since the program was implemented, but they also report that CIQ testing rejected 2 of 58 apple juice concentrate shipments bound for the United States in 2008 for pesticide residues that exceeded tolerances.[10]

Quality and safety strategies call for closer links in the supply chain and adoption of production standards in orchards. Shaanxi's CIQ encouraged juice processors to adopt a "company + production base + standard" supply chain that implies fixed supply relationships with groups of apple farmers that follow the company's standards. The CIQ in western Henan executed an agreement with the municipal government to establish 121 new orchards that employ recordkeeping to facilitate traceability systems for apples used in exported juice. Shaanxi CIQ carried out pilot programs to implement Good Agricultural Practices (GAP) and organic production in apple orchards.

The Henan MOFCOM article emphasizes investments in testing and detection capabilities. A local laboratory was accredited by the national accreditation agency and the CIQ in 2008, and newly imported U.S. testing equipment expanded the number of detectible pesticides to 100. The local laboratory began testing juice shipments for melamine adulteration in 2008 and found no traces in 19 samples of fruit juice.

The article also illustrates how CIQs serve as an intermediary to apprise exporters of international standards. After the CIQ in western Henan learned of a new U.S. FDA import alert about heavy metals (arsenic and lead) in grape and pear juice in March 2009, it gathered information about maximum tolerances, informed exporting companies, and required them to implement internal controls and test products. The CIQ also carried out its own testing of export shipments. It reported that none of the samples tested exceeded tolerances.

CONCLUSION

China will likely continue as the world's leading apple juice concentrate exporter. With China accounting for about half of the world's apple production, it is unlikely that any other country could match China's supply of raw materials. Chinese officials plan further expansion. Shaanxi Province set a target to expand apple production to 10 mmt by 2012 (*Invest in Shaanxi*), up from 8 mmt in 2009. As earlier, officials in western Henan Province launched a project to set up 121 new orchards totaling over 40,000 acres in 2009 (Henan MOFCOM). However, further expansion will require pushing apple production into marginal production areas.

While China's industry supplies a large share of the world's apple juice concentrate, the industry's reliance on price for its competitive position leaves it in a tenuous long-term standing (Li et al.). Chinese juice suppliers have not developed strong brands or consumer product lines, and the industry is predominantly a supplier of low-cost generic ingredients to multinational consumer product companies. This leaves the Chinese industry vulnerable to wide swings in demand resulting from the purchase decisions of price-sensitive downstream users (Chen). The sudden boom in demand for Chinese juice followed by a sharp reversal during 2007-08 illustrates the vulnerability of the industry to shifts in demand for its product. Juice from China competes with apple juice from other countries and other types of juice and ingredients, and it may be affected by swings in consumer preferences toward nonjuice beverages. Appreciation of China's currency may also affect the cost of China's apple juice exports. For example, if the exchange rate of 6.83 Chinese yuan per dollar that prevailed during 2008-10 appreciated 10 percent to 6.15, the cost of 7 metric tons of juice apples costing 4,200 yuan would rise from the equivalent of $615 to $683.[11] Industry reports from China cited currency appreciation as a factor affecting industry competitiveness when the Chinese yuan appreciated roughly 21 percent during 2005-08, yet China's juice exports soared during that period. China's retail prices of fresh-market apples and other foods are far below U.S. prices when converted to U.S. dollars (Gale and Tuan). This analysis of the apple juice industry reveals the nexus between small-scale farmers, food industry investors, and government officials behind China's emergence as an exporter of agricultural products. The rise of the apple juice industry illustrates the mobilization of investment from a combination of private companies, central and local government, and foreign investors to utilize agricultural raw materials to build an industry and acquire a large share of the world market in a remarkably short time. Procuring raw materials from small farmers in China's hinterland to supply ingredients to multinational food and beverage companies presents many challenges in coordinating the volume and quality of raw material supplies with demand from the global market. Other industries, like corn-based industrial products (Gale et al.) and dairy products (Fuller et al.; Gale and Hu), carried out similar rapid expansions driven by investment in industrial processing of agricultural commodities. Most of these industries meet emerging demand by Chinese consumers (for example, dairy products), but many produce goods for export (for example, citric acid produced using corn as a raw material).[12] As China's infrastructure construction and western development campaign push agricultural industrialization deeper into its hinterland, some of the country's other food exports, such as tomato paste, canned mandarins, canned pears and pear juice, and dates, are emerging on the world market following a path similar to that of China's apple juice industry.

REFERENCES

Baidu, Baike. "*Xibu Da Kaifa (Western Development)*," Undated encyclopedia entry, http://baike.baidu.com/view/13834.htm

Barboza, David. "Export Apple of China's Eye Is, er, Apples," *The New York Times*, April 4, 2003.

Beckman, Chanda; Wugang Bu; May, Liu. *Peoples Republic of China Fresh Deciduous Fruit Annual 2008*, U.S. Department of Agriculture, Foreign Agricultural Service, GAIN CH8108, November 15, 2008.

Carter, CA; Gunning-Trant, C. "China's Food Exports Face Antidumping Laws," *American Journal of Agricultural Economics*, Vol. 88, No. 5 (2006), pp. 1227-1234.

CCCFNA (China's National Native Produce and Animal By-products Import and Export Association). "*Qiye Jiti Qicao "Zhongguo Nongsuo Guozhi Qiye Shipin Anquan Beijing Chengnuo Shu*" (Enterprises Collectively Draft "Beijing Food Safety Pledge of Juice Concentrate Enterprises")," December 7, 2007, http://cccfna.mofcom.gov.cn/aarticle/ztxx/200712/20071205275631.html

Chen, Xiaohong. *"Nongsuo Pingguozhi Hangye Kuaisu Kuozhang Li Yihuan (Hidden Danger in the Rapid Expansion of the Apple Juice Concentrate Industry)*," Unpublished industry report, web site for the trade promotion of small and medium enterprises, August 2005, www.tpbjc.gov.cn/Article_Show.asp?ArticleID=6008

China Juice Concentrate Net. "Zhongguo Pingguozhi Shengchan Chengben Jisuan (Calculating Chinese Apple Juice Production Costs)," Unpublished industry report, September 17, 2009, www.fjc100.cn/news/ market info/2009/9/099171359174352.html

China Juice Concentrate Net. "Woguo Nongsuo Pingguozhi Hangye Fazhan Zongshu (Overview of China's Concentrated Apple Juice Industry Development)," Unpublished industry report, March 26, 2009, www.ytxin-gnong.com/html/news-95.shtml

China Ministry of Agriculture. *Apple Advantaged Product Regional Layout Plan* (2008-15), February 2009, http://hiagri.gov.cn/eWebEditor/ upload-file/20090219182428236.doc

China Ministry of Agriculture Development Plan Office. *Compilation of Advantaged Agricultural Product Regional Layout Plans*, Beijing: China Agricultural Press, 2004.

China National Bureau of Statistics. *Rural Statistical Yearbook*, Beijing: China Statistics Press, various years.

China National Bureau of Statistics. *Rural Household Survey Yearbook*, Beijing: China Statistics Press, various years.

China National Development and Reform Commission. *Compilation of Materials on China Agricultural Products Cost and Return*, Beijing: China Statistics Press, 2009.

Cui, Kai; Yifan, Pan. *Industrial Map of Chinese Food*, Beijing: China Light Industry Press, 2006.

Cui, Yue. "Nongsuo Pingguozhi Hangye: Dui Zhiliang Wenti Shuo Bu (Concentrated Apple Juice Industry: Say No to Quality Problems)," *Online International Business Daily,* December 4, 2007.

Economic Reference News. "Raising Fruit Juice International Competitiveness: Need to Overcome Raw Material Shortage," May 20, 2009, http://ccn.mofcom.gov.cn/swxw/show.php?eid=29992

Fryhover, Todd. "Apple Lines—Full Access to China Needed," *Good Fruit Grower*, April 1, 2010.

Fuller, Frank; Huang, J; Ma, H; Rozelle, S. "Got Milk? The Rapid Rise of China's Dairy Sector and its Future Prospects," *Food Policy*, Vol. 31, (2006), pp. 201-215.

Gale, F; Buzby, Jean C. *Imports from China and Food Safety Issues,* U.S. Department of Agriculture, Economic Research Service, Economic Information Bulletin 52, July 2009, www.ers.usda.gov/publications/eib52/

Gale, F; Hu, D. "*Supply Chain Issues in China's Milk Adulteration Incident*," Paper presented at International Association of Agricultural Economics Meeting, Beijing, China, August 2009.

Gale, F; Tuan, F. *China Currency Appreciation Could Boost U.S. Agricultural Exports,* U.S. Department of Agriculture, Economic Research Service, Outlook Report WRS-0703,August 2007, www.ers. usda.gov/publications /wrs0703/

Gale, F; Tuan, F; Wang, X; Cao, Z. *China Is Using More Corn for Industrial Products,* U.S. Department of Agriculture, Economic Research Service, Outlook Report FDS-09K-01, December 2009, www.ers.usda.gov /publications/fds/2009/12dec/fds09k01/

Global Trade Information Services, Inc. *Global Trade Atlas*, Customs statistics database, accessed November 2009.

Haitong Securities. "Yuanliao Duanque Tuoman Hangye Fusu Bufa (Shortage of Raw Materials Slows Industry Recovery)," January 2, 2010, http://bankuai.yuandaont.com/bkcnt/00104140496.htm

Hefling, Kimberly. "U.S. Apple Growers Feel Heat From China," *The New York Times*, June 25, 2007, www.nytimes.com/2007/06/25/business /worldbusiness/25iht-apples.1.6312540.html

Henan MOFCOM. "Yuxi Diqu Shipin Nongchanpin Chukou Diyi Lidu Chengxian Nishi Bei Zeng (Western Henan Food and Agricultural Exports First Quarter Double in Adverse Conditions)," *Online news article*, April 22, 2009, http://henan.mofcom.gov.cn/aarticle/sjtongjizl/200904/20090406190766.html

Huang, Sophia; Fred, Gale. *China's Rising Fruit and Vegetable Exports Challenge U.S. Industries*, U.S. Department of Agriculture, Economic Research Service, Fruit and Vegetables Outlook Report FTS-32001, February 2006, www.ers.usda.gov/publications/fts/2006/02feb/fts32001/

International Finance Corporation (IFC). "Good Practices Reduce Poverty in China," Press release, May 13, 2006, www.ifc.org/ifcext/media.nsf/Content/China_GoodPractices_Reduce_Poverty

Invest in Shaanxi. "Zhuanfang Shaanxi Guoyeju Fujuchang Liu Ling (Interview with Vice Director of the Shaanxi Fruit Bureau, Liu Ling)," June 26, 2009, http://sn.ifeng.com/zhuanti/show.php?itemid=2407

Li, Keyong; Guoqiang Chu, Yanhao Xiong. "Guoji Jinrong Weiji "boji" Shaanxi Pingguozhi Chanye (International Financial Crisis Spreads to Shaanxi Apple Juice Industry)," Xinhua News Agency, December 2, 2008, http://nc.people.com.cn/GB/8450644.html

Liu, Qian. "Review of 2008 Apple Juice Concentrate Industry and 2009 Outlook," Unpublished industry analysis report, March 23, 2009.

Liu, Xuetao. "Shaanxi Pingguo Nongsuo Guozhi Jiagong Lin Yuanliao Duanque Weiji (Shaanxi Apple Juice Concentrate Industry Faces Raw Material Shortage Crisis)," *China Economic Times*, December 2006.

Lohmar, B; Gale, F; Tuan, F; Hansen, J. *China's Ongoing Agricultural Modernization*, U.S. Department of Agriculture, Economic Research Service, Economic Information Bulletin 51, April 2009, www.ers.usda.gov/publications/eib51/

Ma, Doris; Tim, Summers. *Is China's Growth Moving Inland? A Decade of "Develop the West",* Chatham House, Asia Programme Paper: ASP PP 2009/02, October 2009.

Naegely, Stella. "The Great Orchards," *Fruit Grower*, September 1998, pp. 6-7.

Nongbo Net. "Yuanliao Wenti Zhiyue Woguo Pingguo Jiagongye Fazhan (Raw Material Problems Constrain Our Country's Apple Processing Development)," March 22, 2007, www.91fruit.com/2009/0326241002009.shtml

Nongbo Net. "Zhongguo Pingguo Nongsuozhi Jiagongye Weiji Sifu Que Qianjing Guangming (Despite Perilous Crisis, Bright Prospects for

China's Apple Juice Concentrate Industry)," Online report, April 29, 2009, www.litchi.cn/2009/0508/2NMDAwMDAwOTM2NQ.html

Nongbo Net. "Fazhan Gaosuan Zhuanyong Guo Jidi Jianshe (Develop Production Bases for High-Acid Special-Use Fruit)," Undated collection of online articles, http://guoshu.aweb.com.cn/old_zt/1462.html

Pu, Yingyan; Yingkuan, Wang; Tianli, Yue; Chenpeng, Gao; Yahong, Yuan. "Construction of Traceability System for Quality Safety of Apple and Apple Juice," *Transactions of the China Society of Agricultural Engineering*, Vol. 24, Supp. 2 (2008), pp. 289 292 (in Chinese with English abstract).

Robinson-Jacobs, Karen. "Chinese Apple Juice Imports Causing Concern," *Dallas Morning-News*, December 8, 2007, www.dallasnews. com/shared-content/dws/bus/stories/120807dnbusjuice.28107a0.html

Sanmenxia Municipal Government. "Zhunque Youxiaode Xinxi Fuwu Zhu Chukou Qiye Duguo "Weiji" Nanguan (Accurate and Effective Information Service Helps Export Enterprises Get Through "Crisis" Difficulties)," Online news article, April 22, 2009, www.smx.gov.cn/ n57732.aspx

Schumer, Charles. "Schumer Details New Plan To Get Tough On China To Help Struggling Ulster County Apple Farmers," Press Release, May 26, 2004, http://schumer.senate.gov/new_website/record.cfm?id=265335

Shaanxi CIQ (Entry-Exit Inspection and Quarantine Bureau). "Guojia Zhijian Zongju "Zhiliang he Anquan Nian" Huodong Jianbao Di 29 Qi (AQSIQ "Quality and Safety Year" Activity Report No. 29)," Online news article, May 26, 2009, www.sncip.gov.cn.

Shaanxi Provincial Government. "Guanyu Jiakuai yi Pingguo Weizhude Guoye Chanyehua Jianshede Jueding (Decision on Speeding up Construction of Apple-Base Fruit Industry)," Provincial Document No. 14, 2000.

Tie, Min. "Jiji Miandui Zhazhi Guo de Zhang Jia (Actively Confronting the Rise in Juice-Fruit Price)," Northwest Horticulture, December 2007, pp. 35-36.

U.S. Apple Assocation. "Chinese Fresh Apple Imports," Issue papers, August 2008, www.usapple.org/industry/legislative/issuepapers.cfm

U.S. Department of Agriculture. Production, Supply and Distribution. Database, accessed April 2010.

U.S. Department of Agriculture, Economic Research Service (USDA/ERS). Food Availability (Per Capita) Data System, Online database, accessed April 2010, www.ers.usda.gov/data/food consumption/foodavailIndex.htm

Wang, Lansu. "Gansu Qin'an: Zhazhi Pingguo Die Zhi Shen Yuan (Gansu Qin'an Juice Apples into the Abyss)," *Lanzhou Morning News*, November 6, 2008, www.tianshui.com.cn/news/qa/ 2008110610245 965864.html

Warner, Geraldine. "Apple Concentrate Antidumping Order Extended," *Good Fruit Grower*, November 2005.

Zhao, Fuquan. "Wang An he Ta de "Jin Pingguo" (Wang An and His "Golden Apples")," *Dangyuan Ganbu zhi You (Friends of Party Member Offi cials)*, December 2006, www.dygbzy.com/html/DYFC/ 2007-7/11/15_00_06_279.html

Zhao, Guangfei. "Shaanxi Wei Fengchan Pingguo Tuo Xiao Lu (Shaanxi Opens Market for Big Apple Harvest)," *Farmers Daily*, October 13, 2009, http://news.aweb.com.cn/2009/10/14/117200910 140937250.html

Zhou, Suocheng; Yingcai, Guan. "Pingguo Chanye Fazhan Zhongde Wenti Ji Dai Jiejue (Urgent Solutions for Problems in the Development of the Apple Industry)," Linyi Government Net (www.sxly.gov.cn), April 4, 2008.

End Notes

[1] Apple juice antidumping duties, effective in 2000, were waived for some Chinese companies in November 2002; other companies were assessed duties until February 2004. However, the antidumping duties order remained in place after the "sunset" review process in 2005 by the U.S. International Trade Commission and the U.S. Department of Commerce. The law requires that antidumping orders be reassessed every 5 years to ensure that they are still appropriate.

[2] In a 1998 interview, a consumer beverage products company asserted that the abundant supply of foreign concentrate encouraged his company to buy more U.S. juice apples because the supply of raw materials allowed plants to operate at full capacity (Naegely).

[3] The processed share of apple use is 35 percent in the United States and exceeds 50 percent in many European countries.

[4] A number of industry sources (for example, China Juice Concentrate Net) report a 7:1 ratio of apples to juice concentrate.

[5] Note: "Shaanxi" and "Shanxi" are different provinces with similar names.

[6] Preliminary reports say that Shaanxi's output surpassed Shandong's in 2009.

[7] Plans for selected commodities and provinces were formulated for 2003-07 (China Ministry of Agriculture Development Plan Office). A similar set of plans for apples and other commodities were formulated for 2008-15.

[8] The website of the Tongda subsidiary in Shaanxi's Liquan County introduces the company as follows: (author's translation) "Tongda Liquan Fruit Juice LLC…was founded in November 2001 with investment of 78 million yuan from the Shenzhen City Eastern Development Group to imple-ment the central party/government's western development process." http://www.86xc.com/369024.html

[9] Chinese pork, vegetable oil, and dairy prices also increased sharply during this period.

[10] Presumably, the rejected shipments were not exported.

[11] Chinese authorities have indicated a preference for a gradual appreciation of the currency. A 10-percent appreciation might occur over the course of a year or more.

[12] Apple juice concentrate is an unusually export-oriented sector— Chinese consumers traditionally do not consume fruit juices, and over 90 percent of the industry's products are exported. While fruit juice consumption is rising in China, orange juice accounts for most of the increase.

In: Global Agriculture. Volume 1 ISBN: 978-1-62257-369-1
Editors: M.R. Robertson, J.I. Schmidt

Chapter 4

TRADE AND FOOD SECURITY IMPLICATIONS FROM THE INDONESIAN AGRICULTURAL EXPERIENCE*

Nicholas Rada and Anita Regmi

ABSTRACT

As the Indonesian Government transitioned from a food (rice) self-sufficiency focus to an industrial export-oriented development strategy in the mid-1980s, the agricultural sector responded by diversifying production to include cash crops. Sharp currency devaluation and trade liberalization following the Asian financial crisis in 1997 provided incentives for Indonesian agricultural producers to assert their comparative advantage in tropical perennial crops. This report evaluates the ensuing impact on Indonesia's agricultural productivity growth and consumer food demand and lifestyle. Shifting agricultural production and consumption patterns have led to improving agricultural trade patterns and food security. These developments have contributed to increased export opportunities for U.S. agricultural suppliers.

Keywords: Indonesia, agricultural productivity, food security

---

* This is an edited, reformatted and augmented version of the United States Department of Agriculture Publication, WRS-10-01, dated May 2010.

INTRODUCTION

Indonesia's resilient economic growth, expanding middle class, and strong domestic demand have led some economists to see the country as a potential fifth member of the BRIC group—Brazil, Russia, India, and China—of rapidly growing developing nations (Aglionby, 2009). Investments in the Indonesian agricultural research system, including U.S. foreign aid, assisted in establishing agricultural research capacity, raising productivity, and supporting stronger growth. The agricultural sector underwent some realignment in response to policy changes that shifted focus from food crop production to an industrial, export-led development strategy. Coupling this market reorientation with the devaluation of the rupiah and trade liberalization following the Asian financial crisis allowed Indonesian producers to exploit their comparative advantage in tropical perennial crop production. Strong economic growth and vibrant trade have increased consumer demand for higher valued foods and have expanded U.S. export opportunities for fruit, dairy, other packaged food, and feed and fodder.

This report examines the evolution of Indonesia's agricultural sector and the ensuing impact on agricultural productivity growth and consumer food demand and lifestyle. The report further examines how the development of the agricultural sector has affected Indonesian agricultural trade patterns, the country's food security status, and the outlook for these trends. At a time when food security initiatives are being developed to reduce poverty and hunger in some of the poorest nations in the world through growth in agricultural productivity, Indonesia can provide an illustration of the wide-ranging possibilities for improving agricultural productivity and food security.

Background

In the mid-1960s, Indonesia was the world's largest rice importer (Fuglie and Piggott, 2006). From the 1970s until the mid-1980s, the country focused its efforts on reducing its reliance on food imports and directed investments toward increasing food crop (mainly rice) production. Growth in the agricultural sector coincided with the "Green Revolution" as seed-fertilizer technologies and substantial Government subsidies allowed increased production through crop intensification. Bolstered by a buoyant food sector in the 1980s, Government sentiment began shifting toward a new industry-first policy. Indonesia's 1983-93 market transformation restructured the economy to assist producers in competing with low commodity prices observed in world markets (Timmer, 2004).

The Asian financial crisis shocked the Indonesian economy, plunging its 1998 gross domestic product (GDP) growth rate to -13 percent (World Bank, 2008). Regaining macroeconomic stability required instituting specific policy measures, negotiated with the International Monetary Fund (IMF), such as reducing food crop tariffs to no more than 5 percent, deregulating the movement of interprovincial agricultural commodities, and breaking the Government Logistical Agency's (Badan Urusan Logistik) monopoly over the trade rights for rice, wheat, soybeans, sugar, and garlic (USDA, Economic Research Service, 2000; Timmer, 2004). Trade liberalizing policies and currency devaluation promoted a greater market-oriented economy and allowed Indonesia's tropical perennial crops to assert their comparative advantage.[1] By the late 1990s, Indonesia emerged as the second largest exporter of rubber and palm oil and the third largest exporter of cacao and coffee (Fuglie and Piggott, 2006). Despite diversifying away from food crops, the country improved food security, largely by increasing access to food with a revitalized economy and rising income levels. In a sample representing 40 percent of the lowest income group in both rural and urban areas, a United Nations' (UN) analysis found that the proportion of Indonesian households experiencing severe food insecurity decreased from nearly 31 percent in 1999 to just under 11 percent in 2002 (United Nations, 2008, p. 30-2).[2]

Indonesian Agriculture's Shifting Production Patterns

Total agricultural output in Indonesia grew through land expansion, increased labor and capital investments, and intensified use of intermediate inputs, such as fertilizer.[3] Average annual agricultural output growth in Indonesia increased 3.6 percent between 1961 and 2006, with roughly half accounted for by farms increasing input usage (Fuglie, 2009). The remaining portion of output growth is attributed to total factor productivity growth—that is, growth from improved technologies and more efficient allocation of farm resources. While technical changes and improved information may come from various sources, both international and domestic, a country's own agricultural research system bears the burden of improving its farmers' productivity.

Although Indonesian agricultural research dates to the early 1800s, the Great Depression, World War II, the War of Independence (1945-49), and nationalization of many foreign-owned plantations (1957) decimated research capacity. To assist in rebuilding domestic public agricultural research capabilities, the World Bank, the U.S. Agency for International Development (USAID), and the Asian Development Bank have accounted for nearly a third of total annual research expenditures between 1974 and 2003 (figure 1).[4] Despite these international contributions, low funding for agricultural research in Indonesia has historically been a constraint to improving agricultural productivity growth (Agency for Agricultural Research and Development, 1987; Fuglie and Piggott, 2006; Mellor et al., 2003). Private agricultural research is increasing but from a very small base: It accounted for about 3 percent of all agricultural research expenditures in 1985, about 7 percent in 1996, and has continued to grow since (Pray and Fuglie, 2002).

Empirical evidence indicates that research investments have benefited Indonesian agricultural development. Agricultural research investments on rice were estimated to generate a rate of return, which is the ratio of money gained relative to the amount invested, of 151 percent annually between 1965 and 1977 (Salmon, 1991). Similarly, other studies have estimated at least a 100 percent return for dryland rice, wetland rice, maize, soybeans, sweet potatoes, cabbage, potatoes, garlic, mustard, onions, shallots, bananas, papayas, pineapples, and milk (Evenson et al., 1994). However, elevating agricultural total factor productivity (TFP) growth is generally the goal of any public agricultural research investment. TFP provides the most complete productivity measure, compared with partial land or labor productivity measures, because it accounts for all farm inputs, rather than a single input, related to production.

Fuglie's (2009) TFP estimates—which account for crops, livestock, and cultured fisheries—indicate that Indonesia's agricultural productivity increased 2.18 percent per year during the early years of investment when the focus was on food (rice) self-sufficiency. A shift in focus favoring industrial development resulted in a dramatic slowdown of agricultural productivity growth to less than 1 percent between 1985 and 1997. Market liberalization and sharp currency devaluation following the Asian financial crisis promoted the development of export-oriented tropical perennial crops. Subsequently, TFP estimates more than tripled to 2.41 percent between 1998 and 2006.

Analysis by Rada et al. (2010) finds that agricultural technology growth between 1985 and 2005 has varied across subsectors, being greatest in perennial (export) crops (2.20 percent), followed by livestock (1.70 percent), and least in annual (food) crops (0.67 percent). This analysis suggests that policy reforms and currency devaluation created incentives for increased agricultural trade and generated growth in agricultural productivity. Furthermore, Rada et al. indicate that technology growth was driven more from private and other nongovernment sources than from public agricultural research investments. One such example is provided by Pray and Fuglie (2002), who note that new palm oil clones destined for Indonesia originate from Malaysia, with private rather than public research facilitating the technology transfer.

The realignment of the Government's development focus, accompanied by increased private and nongovernment research investments in perennial crops—which experienced the greatest productivity gains—has contributed to changing Indonesia's agricultural composition. Although most crops and livestock experienced growth in total production levels, their relative shares of total agricultural output changed over the years.

Total agricultural output in Indonesia grew from an average level of 93 million metric tons during 1985-1997 to 117.5 million metric tons during1998-2005 (figure 2). The eight commodities presented in figure 2 accounted for over 85 percent of total agricultural production in both periods. Between these two periods, average rice production increased over 13 percent, which is significant in the face of low annual crop technology growth during this timeframe. In contrast, palm oil production—a principle driver of perennial crop technology growth—increased 110 percent between the two periods.

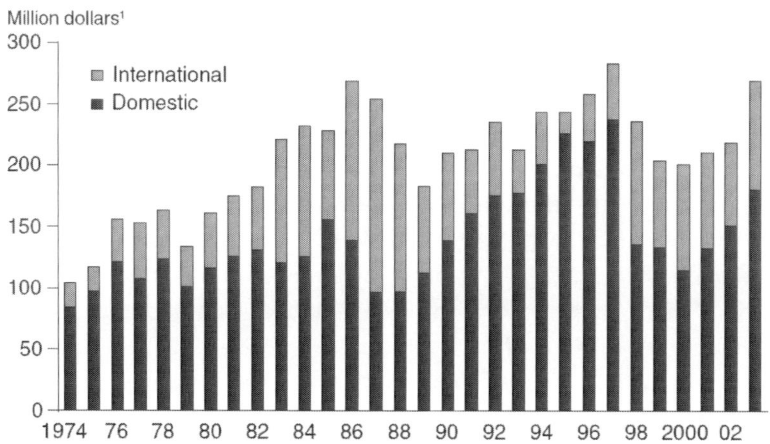

[1] In 1999 international dollars.

Source: Fuglie and Piggott (2006).

Figure 1. Sources of funding for Indonesian agricultural research, 1974-2003.

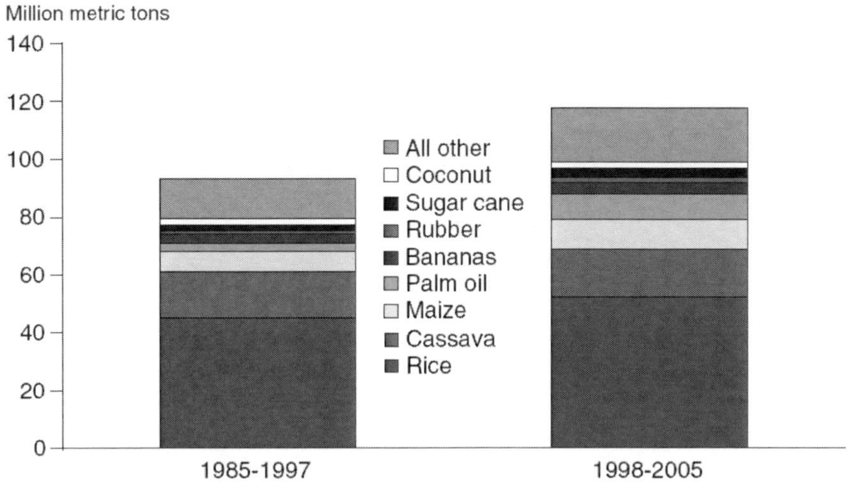

Source: Rada (2009).

Figure 2. Average Indonesian agricultural production, 1985-1997 and 1998-2005.

Just as changes in development focus and market liberalization have impacted Indonesian agricultural productivity trends and production composition, these forces have combined with overall economic growth to drive changes in consumer food demand and lifestyles. These factors are briefly described in the following section.

Agricultural Demand Transitions

Growth and development in the agricultural sector have increased agricultural value added from $17 billion in 1985 to almost $29 billion in 2003, in constant 2000 dollars (World Bank, 2008). However, because the rest of the economy grew even faster, agriculture's share in the general economy has declined by nearly half (table 1). Agriculture accounted for over 30 percent of the country's GDP in 1985 but less than 17 percent in 2003. Agricultural productivity growth reduced labor needs in the sector, promoting greater labor participation in other sectors of the economy. Additionally, higher wages in urban areas have stimulated rural labor migration (World Bank, 2007). As a result, agriculture accounted for only about 44 percent of total employment in 2003 compared with almost 55 percent in 1985 (table 1). Furthermore, the urban population rose to account for nearly 50 percent of the overall population in 2006; it accounted for only 15 percent in 1961 (figure 3).

Economic growth and greater urbanization have contributed to lifestyle changes, evidenced by a more educated labor force with greater female participation (table 1). Increasing employment opportunities have coincided with higher per capita wealth. GDP per capita (in 2000 international dollars) nearly tripled from $1,150 in 1975 to almost $3,200 in 2003 (table 1). Past ERS research indicates that higher income levels are associated with a greater demand for more expensive sources of calories, such as meat, fruit, vegetables, and processed food products (Seale et al., 2003). For example, if Indonesian income levels rose 10 percent, the demand for higher valued foods, such as dairy and fish, would rise by about 8 percent for each, whereas cereal demand would increase by only 5 percent.

Urbanization affects food consumption patterns in a number of ways (Regmi and Dyck, 2001). Calorie requirements of urban and rural residents differ, with sedentary urban lifestyles requiring fewer calories to maintain a given body weight. Improved transportation technology offers a wider range of food products in urban areas year-round than in rural areas. Moreover, urban areas are centers of economic opportunity where a greater percentage of women work outside their homes, in turn generating larger household incomes.

Table 1. Indonesian economic development indicators, 1975-2003[1]

Year	GDP growth	GDP per capita, PPP[2]	Agricultural GDP share	Agricultural employment[3]	Female labor	Literacy rate[4]
	Percent	*Constant 2000 U.S. dollars*	*Percent*	*Percent*	*Thousands*	*Percent*
1975	7.8	1,148	30.2	—	—	68.7
1985	6.0	1,735	23.2	54.7	25,414	76.1
1995	7.1	2,987	17.1	45.2	31,796	81.9
2003	4.7	3,182	16.6	44.3	37,979	85.8

[1] Each indicator is provided as a 5-year centered average to net out year-to-year fluctuations.
[2] Gross domestic product (GDP) per capita, purchasing power parity (PPP) measured in constant 2000 U.S. dollars. 1975 employs a 3-year uncentered average as data begin in 1975.
[3] — = Not applicable. 1985 is the first data point.
[4] 1975 employs only 1976 data. 2003 employs a 3-year average as data end in 2003.
Sources: World Bank (2008), Fuglie and Piggott (2006), Euromonitor (2009).

Greater food consumption and increased purchasing power have contributed to average Indonesian per capita food availability growing from 1,726 calories per capita per day in 1961 to 2,890 calories in 2003 (FAO, 2009). While the increase in total per capita calories has been remarkable, the changes in food sources providing these calories are equally noteworthy. The total calorie share of starchy roots, a low-value product, has been consistently replaced by high-value foods, such as vegetable oils, meats, fish/seafood, and fruits and vegetables (figure 4). Between 1961 and 2003, starchy root crops as a share of total calories consumed per capita per day fell an annual average 2.5 percent, from 20 percent to 7 percent. Cassava, long considered an inferior food in Indonesia, is the dominant starchy root. Vegetable oil's share nearly tripled between 1961 and 2003, peaking at more than 9 percent. Meat and fish/seafood as a share of calories modestly rose 1.1 percent and 1 percent, respectively. Sugars and sweeteners and vegetables and fruits remained constant.

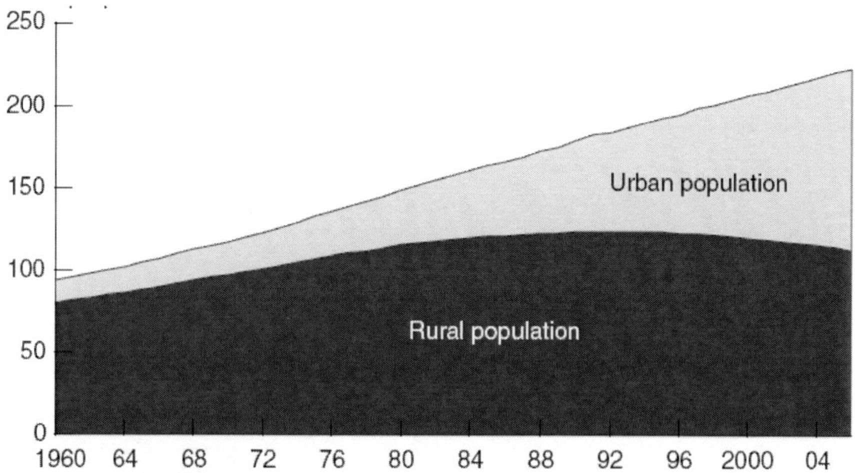

Source: World Bank (2008).

Figure 3. Indonesian population growth, 1960-2006.

Income growth and lifestyle changes caused by greater urbanization have also changed the manner in which consumers purchase food in Indonesia. Increasingly, Indonesians are purchasing packaged food with some value added, rather than purchasing unprocessed products from local wet markets. The total value of such purchases grew from $4.2 billion in 1998 to over $16 billion in 2008 (figure 5). In addition to the changes in the actual consumer food basket, the outlets through which these products are purchased have also undergone a change. In 1998, less than 22 percent of packaged food was sold in standardized retail outlets rather than in independent corner "mom-and-pop" stores.[5] In 2008 over 34 percent of sales were through standardized stores.

These standardized outlets are often owned by such companies as Carrefour and Makro (recently purchased by Lotte), who play a key role in introducing new food products to the market. In this manner, changes in the food retail sector further contribute to the evolving food demand pattern of the Indonesian consumers.

With a surge in the number of dual-income urban households Indonesians are finding less time to shop for food or cook their meals, registering a dramatic growth in sales from the foodservice sector (figure 6). Although the size of the Indonesian market at $29 billion is relatively small compared with that of the United States, it is one of the world's fastest growing markets. As with food retail outlets, a large number of foodservice outlets that have proliferated in recent years are foreign affiliates of companies, such as fast food outlets McDonald's and Pizza Hut, as well as sit-down restaurants, such as TGI Friday's, Tony Roma's, and Planet Hollywood.

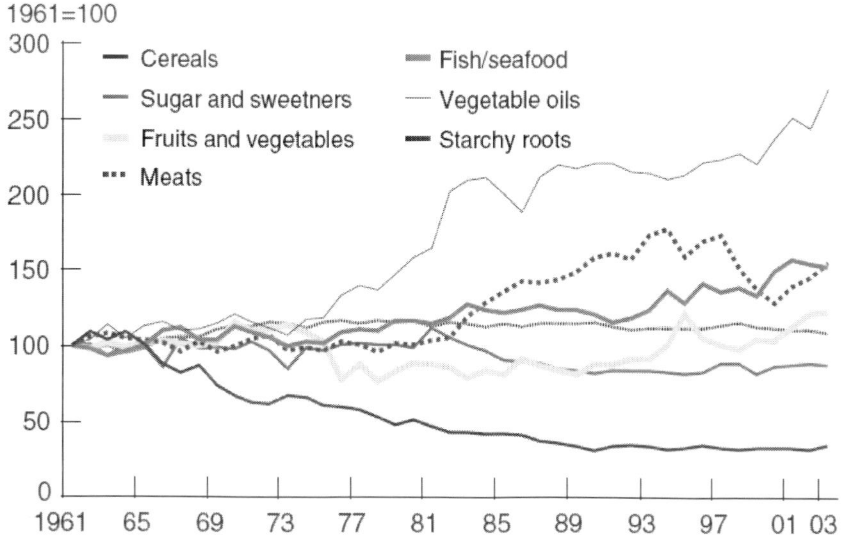

Source: World Bank (2008).

Figure 4. Changes in Indonesian food sources as a share of total calories consumed per capita per day, 1961-2003.

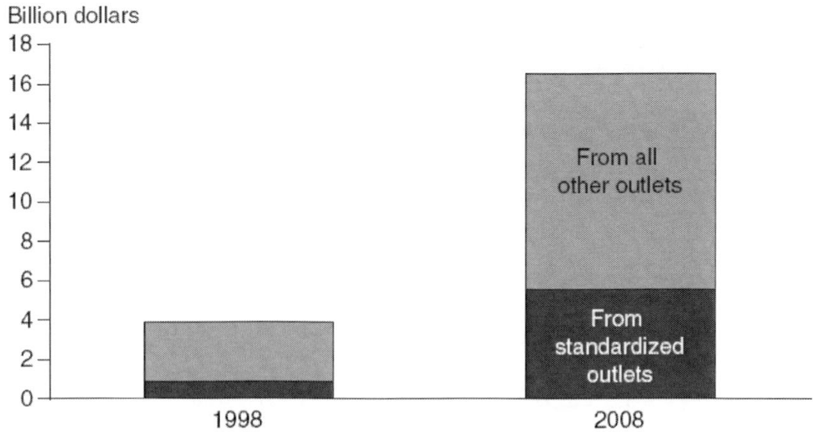

Source: Euromonitor (2009).

Figure 5. Rising packaged food sales in Indonesia, 1998-2008.

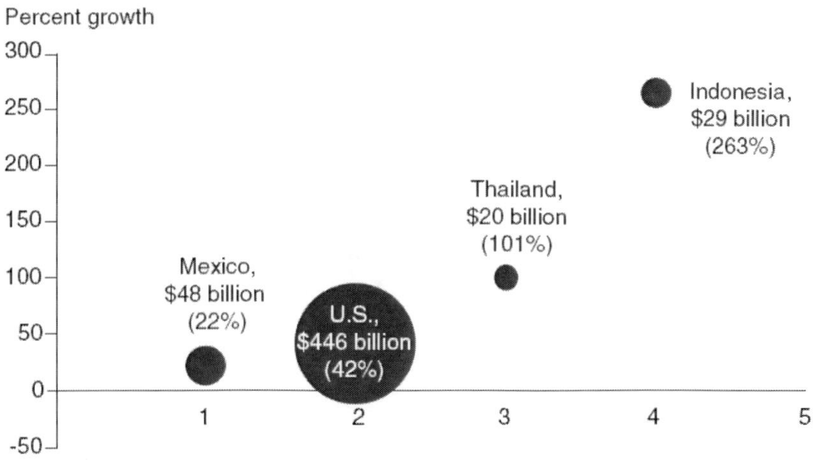

Source: Euromonitor (2009).

Figure 6. Indonesian foodservice market: Size in value, 2008, and percentage growth, 1999-2008.

Food Security

Although Indonesian agriculture has diversified away from food crops, overall food security has improved. Food security generally implies that a country has sufficient food available (target of 2,100 calories per person per day) that can be readily accessed (physically available as well as affordable) and utilized by it citizens (nutrients consumed and absorbed by individuals). Growth in agricultural production has increased food availability in Indonesia, while higher income levels have improved food accessibility for most of its citizens. Increased education, better nutrition, and improved food safety and sanitation have contributed to better food utilization in the country.

With increased availability, accessibility, and better utilization, food insecurity in Indonesia has declined since 1981 (figure 7). Food insecurity results in undernourished citizens, with children being the clearest indicator of this crisis. Thus, food security measures commonly tend to focus on indicators involving children. The International Food Policy Research Institute's (IFPRI) global hunger index provides a measure that reflects all three pillars of food security: availability, accessibility, and utilization. The hunger index is an equally weighted index of three measures: the proportion of undernourished population as a percentage of the total population; the prevalence of underweight children under the age of 5; and the under-5 mortality rate (IFPRI, 2006). The hunger index score for Indonesia has declined from a high of over 28 in 1981 to less than 12 in 2007 (IFPRI, 2006, 2007). If one assumes food availability, agricultural trade, and economic development to continue their historical trends, the hunger index is expected to decline to 8 in 2010 and below 2 by 2020.

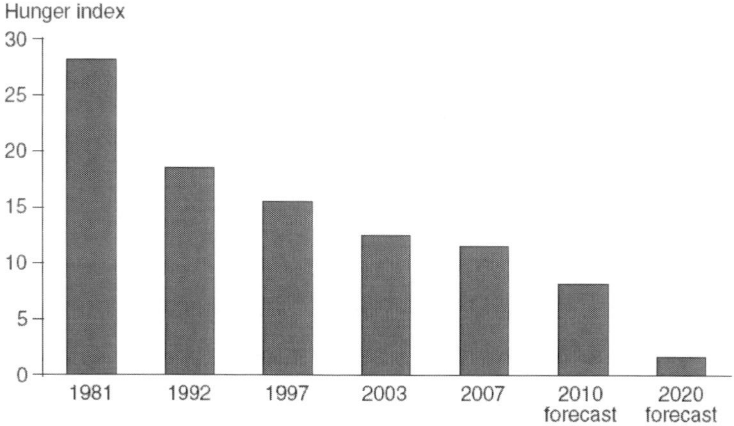

Source: The hunger index is from the International Food Policy Research Institute; values for 2010 and 2020 are authors' trend line forecast.

Figure 7. Improving food security in Indonesia, 1981-2020.

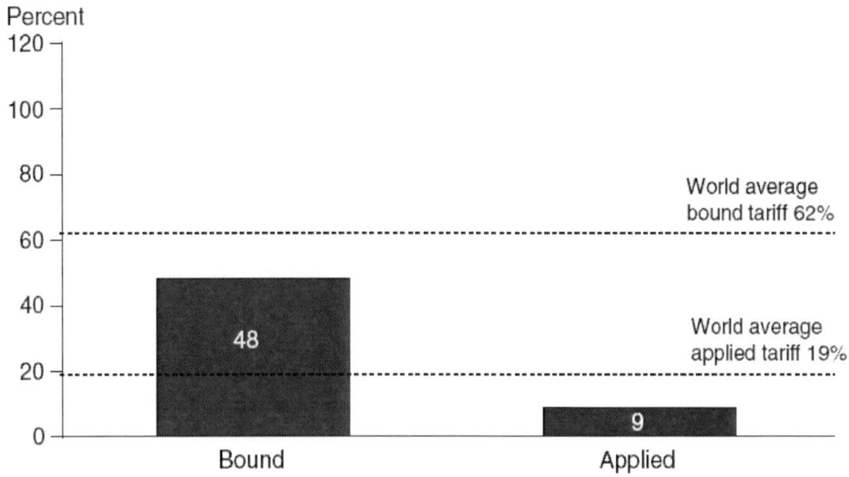

Source: Calculations by USDA, ERS using the Agricultural Market Access Database and the World Trade Organization's member-submitted ad valorem-equivalent estimates; simple averages are calculated.

Figure 8. Indonesia's agricultural tariffs, 2006.

Marginal growth in annual crop productivity suggests that imports may have played a key role in improving consumer availability and accessibility to food. The following section examines how transitions in Indonesia's agricultural production patterns and consumer food demand patterns have affected agricultural trade.

AGRICULTURAL TRADE

With import tariffs below world average levels, Indonesia has one of the most open markets for agricultural imports in Asia (figure 8). For example, in 2006, the average bound rate, a set tariff level above which import tariffs cannot be raised, was 48 percent in Indonesia compared with the global average of 62 percent. More importantly, the actual tariffs that were applied to agricultural imports averaged only 9 percent in Indonesia compared with the world average of 19 percent.

Reflecting the changes in agricultural production and consumer demand patterns, Indonesia has focused agricultural exports primarily on tropical perennial products in which it has comparative advantage, whereas its imports have included feed for its growing poultry sector (in response to greater consumer demand for meat) and food for its citizens.

Agricultural Exports

Indonesia's agricultural export value has grown on average almost 9 percent annually, from a base of nearly $900 million in 1975 to nearly $18 billion in 2007 (FAO, 2009). Growth has been driven by increases in tropical perennial crops, such as rubber, cocoa, coffee, and palm oil. As of 2008, Indonesia was the second largest exporter of palm oil and the fourth largest exporter of coffee (USDA, 2008). While growth was evident in palm oil and rubber following Indonesia's move toward industrialization in the mid-1980s, growth was exponential following the economy's recovery after the Asian financial crisis (figure 9). Between 1975 and 2007, palm oil and rubber accounted for nearly half of total export value, with rubber's share alone nearing a third. Between 2000 and 2007, shares associated with palm oil and rubber switched, with palm oil representing nearly 33 percent of total export value.

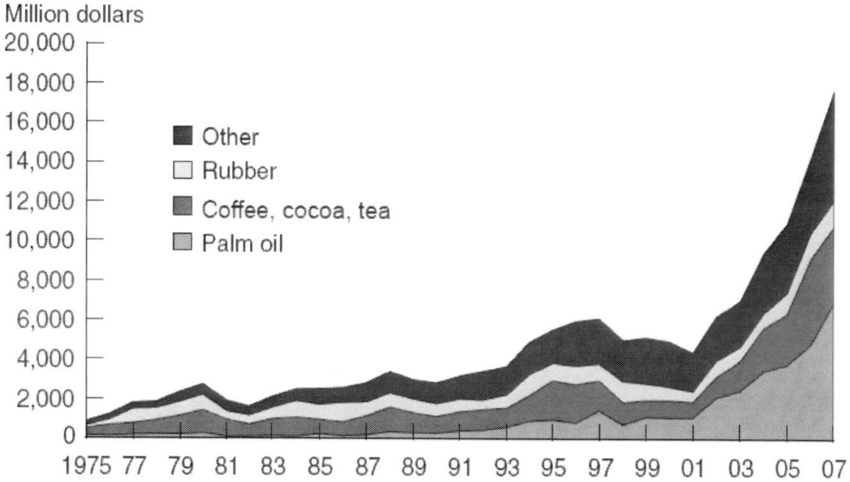

Source: Food and Agriculture Organization of the United Nations, FAOSTAT Agricultural Databases (2009).

Figure 9. Indonesian agricultural export values, 1975-2007.

Agricultural Imports

Despite growth in agricultural production, population and income growth have contributed to Indonesia's agricultural import increases (figure 10). Indonesia's agricultural import value grew from over $650 million in 1975 to over $8.5 billion in 2007, an 8-percent average annual increase, to meet the needs of a swelling population that increased from less than 100 million in 1961 to nearly 230 million in 2009 (FAO, 2009). In addition to the overall growth in imports to meet expanding food demand, Indonesian food import trends reflect food preferences and lifestyle changes of an increasingly educated, urban, and wealthy consumer. Cereals, which accounted for over 60 percent of total imports in 1975, have accounted for only about a quarter of agricultural imports since 1991 (figure 11). Instead, imports of feed and fodder to meet the needs of the growing poultry sector and foods to satisfy Indonesia's newly discerning consumers have risen. For example, feed and fodder import value increased an average 34 percent each year between 1967 and 2007. Similarly, soybean (food item) import value increased an average 12 percent per year between 1975 and 2007.

U.S. Agricultural Exports to Indonesia

Indonesia's agricultural transformation has provided opportunity for U.S. agricultural exports. Since 1975, grain and soybeans have averaged 50 percent of all U.S. agricultural exports to Indonesia (figure 12) (U.S. Census Bureau, 2009). Most noticeable in figure 12 is the sharp growth starting just before 2000. To provide more detail of which U.S. exports have grown since 1997, table 2 displays the increasing values of bulk, intermediate, and consumer-oriented agricultural items exported to Indonesia between 1998 and 2008. Although the value of bulk items far exceeded that of intermediate or consumer-oriented items, each category experienced impressive growth. Furthermore, the average annual value of at least six individual items—wheat, hide and skins, feed and fodder, snack foods, dairy products, and fresh fruits—grew more than 20 percent. Dairy products, the fastest growing export item to Indonesia, grew at an annual average rate of almost 40 percent.

Increased U.S. export values after 1998 reflect changes in the composition of trade (table 2). Export volumes follow value trends for most commodities/groups. The significant average annual decline in "other bulk items" is largely explained by a nearly 40-percent average annual decline in rice quantities between 1998 and 2008. In contrast, strong average annual growth of U.S. export volumes to Indonesia in feed and fodder (28 percent), dairy products (31 percent), and snack foods (20.5 percent) exemplifies the Indonesian consumer's rising demand for animal feed and high-value food items.

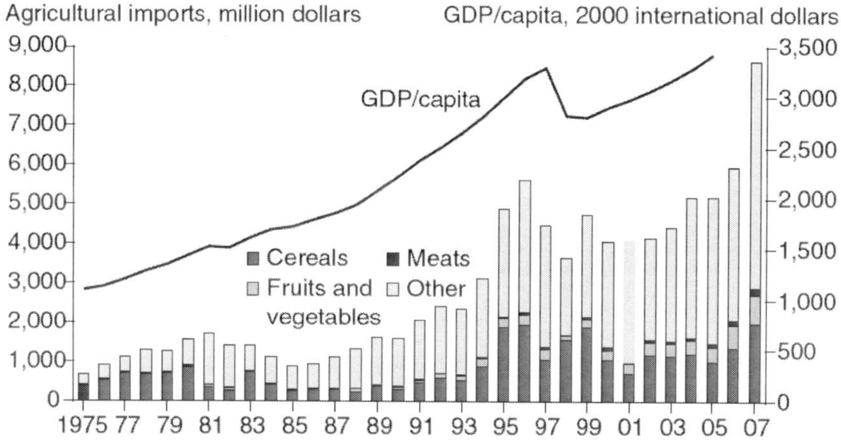

Source: Food and Agriculture Organization of the United Nations, FAOSTAT Agricultural Databases (2009); World Bank (2008).

Figure 10. Indonesian income growth and agricultural import value, 1975-2007.

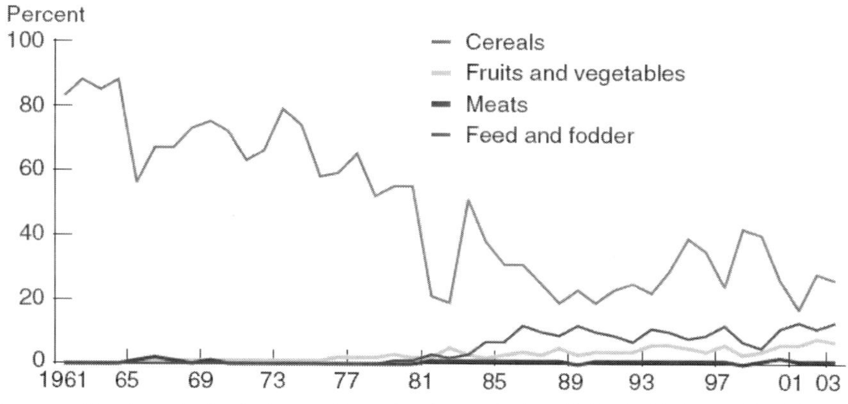

Source: Food and Agriculture Organization of the United Nations, FAOSTAT Agricultural Databases (2009).

Figure 11. Selected agricultural products as a share of total Indonesian imports, 1961-2007.

Table 2. United States-to-Indonesia agricultural export growth, by commodity, 1998-2008

Commodity	1998	2001	2005	2008	Average annual growth,1998-2008	
					Value	Volume
	Million dollars				*Percent*	
Bulk:						
Wheat	44.9	72.4	26.8	364.0	20.9	10.2
Cotton	154.6	194.0	277.6	455.0	10.8	10.7
Soybeans	139.2	244.7	302.2	589.3	14.4	7.9
Coarse grains	1.8	43.8	1.2	10.4	17.8	7.5
Other	33.2	19.2	27.0	43.8	2.8	-17.3[1]
Total	373.6	574.0	634.7	1,462.0	13.6	8.3
Intermediate:						
Hide and skins	0.2	4.0	3.2	3.0	28.4	N/A[2]
Soybean meal	11.7	167.5	22.5	20.8	14.7	-1.5
Feed and fodder	9.7	21.4	72.8	189.1	29.7	28.2
Other	34.6	61.0	55.4	138.6	13.9	N/A[3]
Total	56.2	253.9	153.8	381.5	19.2	N/A
Consumer-oriented:						
Snack foods	0.7	1.7	9.3	6.3	22.1	20.5
Dairy products	4.2	14.5	61.4	209.2	39.0	30.8
Fresh fruit	4.9	29.5	37.6	59.6	24.9	17.1
Meats	3.8	10.9	19.1	9.8	9.4	10.1
Processed fruits and vegetables	9.2	11.6	16.4	27.5	11.0	6.5
Tree nuts	0.4	1.0	2.2	3.0	19.5	19.7
Pet foods	0.5	1.9	1.8	3.3	19.4	12.6
Other	5.1	7.8	21.8	32.8	18.5	N/A
Total	28.9	78.9	169.5	351.5	25.0	N/A
Total agricultural exports	495.4	949.5	1,025.3	2,268.3	15.2	N/A

[1] Excludes peanuts and pulses due to lack of data availability.

[2] N/A = Not available.

[3] Due to differing units of measurement, volume aggregates are available only for bulk items.

Source: U.S. Census Bureau (2009).

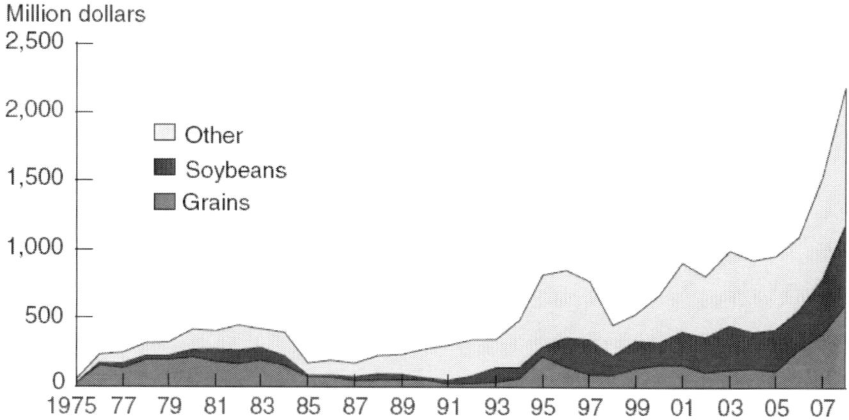

Source: U.S. Census Bureau (2009).

Figure 12. U.S. agricultural exports to Indonesia, 1975-2008.

Whereas figure 12 illustrates the high share that grains and soybeans represent in U.S. exports, table 2 traces new growth to specific agricultural items. Moreover, the growth in export quantities indicated in table 2 confirms that rising U.S. export values are not completely price driven. Indonesia's rising demand for high-value foods has provided a boon for U.S. agricultural exports.

OUTLOOK AND IMPLICATIONS

Indonesia has recorded one of the fastest agricultural transformations in history, trailing only Korea and Turkey, in shedding 27 percent of agriculture's share of GDP in 35 years (World Bank, 2008).[6] Incentives induced through the market reorientation of the mid-1980s and the currency devaluation and market liberalizing policies after the Asian financial crisis may have been the main productivity driver. With greater income, Indonesians are increasingly urbanized and gains have been made in improving their food security. These consumers increasingly demand higher valued food and have attracted multinational stores as well as U.S. agricultural exporters, retailers, and foodservice outlets.

Under the expectation that population and income trends will continue upward, Euromonitor International forecasts that average consumer food expenditures per capita will increase 7.5 percent per year over the next decade, with average packaged food sales growing over 9 percent per year between 2009 and 2014. These forecasts bode favorably for U.S. agricultural producers, who can expect their growing export trends to continue, particularly for poultry feed, dairy, and other high-value food products.

While this historical account of Indonesia's agricultural development and transformation provided the platform for its trade and food security outlook, future expectations are contingent upon the continuation of the policies that thrust it into its present situation. The Indonesian experience provides some insight to donors who are contemplating investments to increase agricultural productivity in less-developed countries in order to reduce global poverty and hunger. Indonesia provides an example of a country in which improvements in food security are not inconsistent with export crops leading agricultural productivity growth. As such, it is important to consider any country's particular geoclimatic characteristics and agricultural comparative advantages before predetermining its food security strategy. Indonesia also indicates that increasing productivity, and thereby food availability, is not sufficient for improving food security; food accessibility also has to be increased by raising the purchasing power of citizens and allowing greater mobility of goods and services.

REFERENCES

Agency for Agricultural Research and Development. (1987). *5 Years of Agricultural Research (1981-1986)*, Jakarta.

Aglionby, J. (2009). "Economists split on Jakarta's Komodo power," *Financial Times*, September 10.

Euromonitor. (2009). *Euromonitor International*, Accessed August, available at http://www.euromonitor.com/.

Evenson, R., Abdurachman, E., Hutabarat, B. & Tubagus, A. C. (1994). "Economic Impacts of Agricultural Research in Indonesia," *Economic Growth Center*, Yale University.

Food and Agriculture Organization of the United Nations (FAO). (2009). FAOSTAT Agricultural Databases, Rome.

Fuglie, K. O. (2009). "Sources of Growth in Indonesian Agriculture," *Journal of Productivity Analysis, Online First,* September 16.

Fuglie, K. O. & Piggott, R. R. (2006). *Indonesia: Coping with Economic and Political Instability*, P.G. Pardey, J.M. Alston, and R.R. Piggott (eds.),

Washington, DC: International Food Policy Research Institute, pp. 65-104.

International Food Policy Research Institute (IFPRI). (2006). *The Challenge of Hunger*, Ordering No. 460-9329, Bonn.

International Food Policy Research Institute (IFPRI). (2007). *The Challenge of Hunger 2007*, Bonn.

Kim, H. & Lee, Y. (2003). "Agricultural Policy Reform and Structural Adjustment: Historical Evidence from Korean Experience," *Policy Reform and Adjustment Workshop*, October 23-25, Imperial College London, Wye Campus.

Mellor, J., Falcon, W., Taylor, D., Arifin, B., Said, E. & Pasandaran, E. (2003). *Agricultural Sector Review, Indonesia*, U.S. Agency for International Development.

Organization for Economic Co-operation and Development. (2009). *Agricultural Policy Design in Developing Countries: Interim Report*, Working Paper No. 32, Paris.

Pray, C. & Fuglie, K. O. (2002). *Private Investments in Agricultural Research and International Technology Transfer in Asia*, Published jointly by International Potato Center, Rutgers University, and the U.S. Department of Agriculture, Economic Research Service.

Rada, N. (2009). "International Comparison and Evaluation of Agricultural Productivity Growth," Ph.D. Thesis, Agricultural and Resource Economics, Oregon State University.

Rada, N., Buccola, S. & Fuglie, K. (2010). "Uncovering Productivity Growth in the Disaggregate: Indonesia's Dueling Agricultural Sub-Sectors," Agricultural and Applied Economics Association, Canadian Agricultural Economics Society, and Western Agricultural Economics Association Joint Annual Meeting, Denver, CO, July 25-27, 2010.

Regmi, A. & Dyck, J. (2001). "Effects of Urbanization on Global Food Demand," *Changing Structure of Global Food Consumption and Trade*, A. Regmi (ed.), WRS-01-1, pp. 23-30, U.S. Department of Agriculture, Economic Research Service, http://www.ers.usda.gov/publications/wrs011/wrs011e.pdf

Salmon, D. (1991). *Rice Productivity and the Returns to Rice Research in Indonesia*, R. Evenson and C. Pray (ed.), Ithaca, NY: Cornell University Press.

Seale, J., Regmi, A. & Bernstein, J. (2003). *International Evidence on Food Consumption Patterns*, Technical Bulletin No. 1904, U.S. Department of

Agriculture, Economic Research Service, http://www.ers.usda.gov /publications/tb1904/tb1904.pdf

Timmer, C. P. (2004). *The Road to Pro-Poor Growth: The Indonesian Experience in Regional Perspective*, Working Paper Number 38, Center for Global Development.

U.S. Census Bureau. (2009). Foreign Trade Statistics, Compiled by the U.S. Department of Agriculture, Economic Research Service, Accessed in December 2009.

United Nations Economic and Social Commission for Asia and the Pacific. (2008). *The Impact of Support for Imports on Food Security in Indonesia,* Working Paper Number 101.

U.S. Department of Agriculture (USDA), Economic Research Service. 2000. *International Crises and Agriculture*, WRS-99-3.

U.S. Department of Agriculture (USDA), Foreign Agricultural Service. 2008. Production, Supply and Distribution Database.

World Bank. (2007). *Horticultural Producers and Supermarket Development in Indonesia*, Working Paper 38543.

World Bank. (2008). World Development Indicators On-Line, Accessed January 2010, available at http://publications.worldbank.org/subscriptions/WDI/WDI.html.

End Notes

[1] Note that Indonesia, as of January 2008, altered its import duty on rice to 450 rupiah/kilogram.

[2] Household food insecurity is defined in this UN analysis as a food expenditure share greater than 60 percent and energy intake less than 80 percent of the standard requirement of 2,100 kilocalories/capita/day.

[3] Secondary data employed in this report are from Indonesian Government statistical publications and non-Indonesian sources. The Indonesian data sources are the Buro Pusat Statistik (BPS), or the Central Statistics Office; the Ministry of Agriculture (MOA); and the Director General of Estate Crops, or DG Estate. Non-Indonesian sources are the Food and Agricultural Organization FAOSTAT Agricultural Databases, the World Bank's World Development Indicators, the U.S. Census Bureau, and USDA's Economic Research Service.

[4] Indonesia's domestic public agricultural research expenditures entail funding sources from the Agency for Agricultural Research and Development (AARD) and the Indonesian Planters Association for Research and Development (IPARD).

[5] Standardized outlets denote supermarkets, hypermarkets, and discount and convenience stores.

[6] Indonesia decreased agriculture's share of GDP between 1971 and 2006 from 40 percent to 13 percent, whereas Korea only required 19 years and Turkey 31 years. These statistics are adapted from Kim and Lee (2003) and Organization for Economic Co-operation and Development (2009).

INDEX

A

access, vii, 1, 5, 6, 49, 59, 60, 61, 62, 72, 84, 97, 98, 101, 103, 119
accessibility, 127, 128, 134
accounting, 43, 92, 108
accreditation, 108
acid, 105, 109
acidity, 91, 105
acquisitions, 62
age, 54, 127
aggregation, 23
agricultural exports, 4, 11, 87, 88, 102, 128, 130, 132, 133
agricultural producers, viii, 59, 117, 134
agricultural sector, viii, 5, 117, 118, 122
agricultural suppliers, viii, 117
agriculture, vii, 8, 30, 31, 37, 85, 98, 122, 126, 133, 136
Algeria, 75
Angola, 75
annual rate, 2, 4, 67
antidumping, 89, 110, 114
apples, viii, 87, 88, 89, 91, 92, 93, 94, 95, 98, 100, 101, 102, 104, 105, 106, 107, 108, 109, 111, 114
architects, 43
Argentina, 22, 74, 76, 90, 92
arsenic, 108
Asia, 33, 112, 128, 135, 136
asparagus, vii, 1, 2, 4, 7, 8, 10, 11, 12, 13, 14, 15, 17, 18, 19, 21, 28, 29, 32, 33, 35, 36
assessment, 3, 84
assets, 35
audits, 106
authorities, 98, 115
authority, 107
autonomy, 36
average costs, 64

B

bacteria, 106
banking, 60
banks, 96
barriers, 7, 13, 40, 65
barriers to entry, 65
base, 7, 64, 65, 78, 107, 120, 129
Beijing, 110, 111
Belgium, 74
beneficiaries, 62, 64
benefits, 5, 7, 10, 24, 31, 70
beverages, 109
body weight, 123
Bolivia, 6, 74
Brazil, 22, 24, 75, 76, 90, 118
breeding, 102
Bureau of Labor Statistics, 47, 82
Burma, 46, 72, 74

Burundi, 76
business environment, 2, 6, 32
businesses, 96, 102
buyer, viii, 87, 88
buyers, 106

C

cabbage, 120
cacao, 119
calorie, 124
Cameroon, 76
campaigns, 61
carbohydrate, 70
carbohydrates, 40, 67
Caribbean, 54, 82, 84
case studies, 10
case study, 3
cash, viii, 104, 117
cash crops, viii, 117
census, 44, 45, 49, 50, 52, 55, 74, 81, 82, 83, 130, 132, 133, 136
central bank, 35
challenges, vii, viii, 1, 3, 28, 33, 60, 87, 88, 109
chemical, 107
chicken, 67
children, 127
Chile, 20, 21, 22, 23, 24, 25, 28, 29, 90, 92
China, vii, viii, 2, 4, 5, 7, 8, 11, 12, 14, 15, 22, 25, 26, 27, 33, 46, 72, 74, 75, 76, 87, 88, 89, 90, 91, 92, 93, 94, 95, 96, 97, 98, 99, 100, 101, 102, 103, 104, 105, 106, 107, 108, 110, 111, 112, 113, 114, 115, 118
chinese government, 97, 101
cities, 103
citizens, 127, 134
city, 47, 48, 97, 114
classes, 40, 41, 43, 48, 50, 57, 77
climate, vii, 1, 2, 5, 6, 8, 9, 22, 32, 49, 50, 98
climate change, 8
coastal region, 8, 22, 30
cocoa, 129
coffee, 4, 11, 119, 129
collateral, 85
Colombia, 6, 26, 74
color, 93, 106
commercial, 11, 40, 51, 52, 54, 61, 76, 84
commercial bank, 61
commodity, 2, 6, 10, 20, 29, 36, 58, 85, 104, 119, 132
communities, 19, 67
community, 7, 68
community service, 7
comparative advantage, viii, 117, 118, 119, 128, 134
competition, 4, 10, 14, 15, 18, 20, 21, 24, 29, 43, 46, 72, 98, 104
competitive advantage, 88, 92
competitiveness, 28, 32, 48, 54, 56, 59, 61, 71, 94, 109
competitors, 7, 20, 24
compilation, 73
compliance, 13
composition, 121, 122, 130
conflict, 37
conflict of interest, 37
congress, 78, 82
conservation, 65
consolidation, 54
construction, 97, 102, 109
consumer markets, 105
consumer price index (CPI), 47, 48, 82
consumers, vii, viii, 1, 10, 15, 21, 24, 39, 42, 48, 49, 72, 84, 92, 106, 109, 115, 124, 130, 133
consumption, viii, 2, 3, 15, 18, 20, 39, 40, 42, 43, 54, 57, 60, 61, 67, 68, 70, 71, 72, 85, 91, 106, 113, 115, 117, 123
consumption patterns, viii, 117, 123
coordination, 61
copper, 36
correlation, 68
correlation coefficient, 68
cost, 4, 14, 15, 24, 28, 29, 31, 46, 62, 65, 88, 89, 91, 105, 108
Costa Rica, 29, 75
costs of production, 14, 64

cotton, 59, 77, 85
country of origin, 21, 27
crop, 4, 19, 36, 49, 53, 56, 59, 61, 65, 66, 72, 78, 85, 93, 95, 118, 119, 121, 128
crop insurance, 65, 66
crop production, 49, 118
crops, viii, 8, 14, 19, 22, 28, 32, 54, 56, 61, 64, 65, 84, 85, 98, 101, 117, 119, 120, 121, 124, 126, 129, 134
Cuba, 75
currency, viii, 109, 115, 117, 119, 120, 121, 133
customers, 104, 105, 106, 107
cycles, 30, 46, 52, 53, 58, 59, 60, 63, 84

D

danger, 14
data availability, 132
database, 34, 84, 95, 111, 113
deficiency, 64
Department of Agriculture, 1, 39, 42, 78, 83, 87, 117, 136
Department of Health and Human Services, 70, 71, 82
depth, 2
deregulation, 6
detection, 108
devaluation, viii, 117, 118, 119, 120, 121, 133
developed countries, 134
developing countries, 84
developing nations, 118
diet, 54, 67
Dietary Guidelines, 70, 82
Dietary Guidelines for Americans, 70, 82
direct payment, 64
disaster, 65
discrimination, 6
displacement, 90
distribution, 30, 32
diversification, 28
diversity, 40
domestic demand, 118
dominance, 26
donors, 134
drought, 56
duty free, 14, 23
duty-free access, 6, 43, 72
duty-free treatment, 83

E

earnings, 67
economic development, 102, 123, 127
economic downturn, 36
economic growth, 3, 118, 122
Economic Research Service (ERS), 1, 9, 14, 17, 24, 35, 47, 70, 71, 77, 79, 80, 81, 82, 83, 85, 90, 91, 93, 94, 95, 96, 99, 100, 101, 103, 106, 111, 112, 113, 119, 135, 136
Ecuador, 6
education, 96, 127
Egypt, 74
El Niño, 8
electricity, 50
employees, 7
employment, 8, 19, 122, 123
employment opportunities, 123
energy, 6, 29, 136
entrepreneurs, 6
entrepreneurship, 13
environment, 6
environmental impact, 61
equilibrium, 65
equipment, 62, 95, 106, 108
Europe, 4, 22, 33, 89, 104, 105
European market, 14
European Union (EU), 2, 5, 7, 12, 14, 23, 46, 89, 90, 92
evidence, 120
evolution, 118
exchange rate, 47, 94, 106, 109
expenditures, 67, 68, 69, 120, 133, 136
export market, 3, 4, 7, 21, 22, 25, 28
export promotion, 65
exporter, vii, 1, 2, 8, 10, 14, 22, 24, 28, 32, 88, 89, 90, 99, 108, 109, 119, 129

exporters, 6, 8, 12, 22, 30, 72, 74, 89, 96, 107, 108, 133
exports, vii, viii, 1, 2, 3, 4, 5, 6, 10, 11, 12, 13, 14, 17, 18, 19, 20, 21, 22, 23, 24, 25, 26, 29, 32, 33, 41, 42, 43, 45, 46, 47, 48, 56, 74, 75, 84, 87, 88, 89, 90, 92, 93, 94, 95, 96, 99, 100, 101, 102, 103, 104, 105, 106, 107, 109, 130, 133
exposure, 33
extraction, 30

F

factories, 104
families, 7
Farm Bill, 80
farm size, 32
farmers, 8, 40, 50, 52, 54, 56, 59, 60, 61, 64, 84, 89, 92, 94, 96, 98, 102, 103, 104, 105, 107, 109, 119
farms, vii, 11, 32, 39, 52, 53, 55, 119
FAS, 22, 23, 31
fast food, 125
fat, viii, 39, 70
FDA, 108
Federal Government, 64, 65
field crops, 58, 92
financial, viii, 36, 60, 96, 117, 118, 119, 120, 129, 133
financial crisis, viii, 117, 118, 119, 120, 129, 133
financial intermediaries, 60
financial resources, 96
fish, 120, 123, 124
fishing, 6
fixed costs, 50
flexibility, 65
floods, 8
fluctuations, 123
food, vii, viii, 44, 65, 67, 70, 78, 85, 88, 98, 102, 106, 109, 113, 117, 118, 119, 120, 122, 123, 124, 125, 126, 127, 128, 129, 130, 131, 133, 134, 136
food industry, 109
food products, 88, 123, 125, 134
food safety, 102, 106, 127
food security, vii, viii, 117, 118, 119, 126, 127, 133, 134
Food, Conservation, and Energy Act of 2008, 18, 64
force, 89
Ford, 79, 83
foreign aid, 118
foreign direct investment (FDI), 3, 6
foreign investment, 30, 88, 96, 97
France, 12, 20, 75
free trade, 6, 7, 33, 40, 71
freezing, 14
frost, 8, 56
fruits, vii, 2, 4, 5, 8, 18, 65, 124, 130, 132
full capacity, 114
fundamental forces, 40
funding, 31, 120, 121, 136
funds, 31, 32, 62, 97

G

GDP per capita, 68, 69, 123
Germany, 89
global competition, 23
goods and services, 3, 134
governments, 61, 84, 101
Great Depression, 120
Green Revolution, 118
gross domestic product (GDP), 3, 35, 68, 69, 119, 122, 123, 133, 136
groundwater, 30
growth, vii, viii, 2, 3, 4, 5, 22, 36, 68, 69, 70, 72, 87, 88, 89, 92, 98, 102, 104, 105, 118, 119, 120, 121, 123, 124, 125, 126, 128, 129, 130, 131, 132, 133
growth rate, 4, 69, 119
Guatemala, 26

H

harmful effects, 89
harvesting, 36
health, 7, 25, 70

health care, 7
heavy metals, 108
height, 22
highlands, 7
history, 65, 133
holding company, 97
homes, 123
Hong Kong, 22, 25, 26, 27, 96, 97
horticultural crops, 2
host, 36
hotels, 102
house, 112
household income, 123
housing, 102
humidity, 56

I

import restrictions, 40
imported products, 48
imports, 2, 14, 15, 17, 18, 20, 21, 23, 24, 25, 27, 36, 41, 42, 43, 45, 47, 71, 73, 75, 83, 84, 89, 90, 91, 92, 118, 128, 130, 131
improvements, 71, 134
income, 3, 7, 8, 19, 59, 64, 65, 67, 68, 71, 72, 85, 119, 123, 125, 127, 130, 131, 133
income distribution, 7, 59
income support, 59, 64, 65, 71
independence, 120
India, 7, 75, 76, 118
individuals, 127
Indonesia, viii, 26, 74, 75, 76, 117, 118, 119, 120, 121, 124, 126, 127, 128, 129, 130, 131, 132, 133, 134, 135, 136
industrial processing, 109
industrialization, 109, 129
industries, 2, 6, 8, 9, 10, 18, 33, 35, 40, 101, 102, 109
industry, vii, viii, 1, 2, 5, 8, 9, 10, 11, 12, 13, 14, 15, 18, 20, 22, 23, 25, 28, 29, 32, 66, 87, 88, 89, 91, 92, 96, 97, 98, 101, 102, 103, 104, 105, 106, 108, 110, 112, 113, 114, 115, 118
infestations, 60
inflation, 5
infrastructure, vii, 1, 2, 3, 4, 28, 29, 30, 33, 62, 101, 102, 109
ingredients, 91, 105, 108
insecurity, 119, 127, 136
inspections, 107
institutions, 31, 60
integration, 48
intermediaries, 96, 104
internal controls, 108
international investment, 5
International Monetary Fund (IMF), 119
international standards, 108
international trade, 2, 5
investment, 2, 6, 30, 31, 62, 87, 88, 95, 97, 98, 99, 101, 102, 104, 109, 114, 120
investment bank, 97
investments, 4, 29, 32, 62, 88, 95, 97, 108, 118, 119, 120, 121, 134
investors, 5, 6, 32, 97, 109
Iran, 76
irrigation, 8, 11, 30, 31, 32, 49, 50, 51, 62, 84
issues, vii, 3, 6, 13, 14, 20, 33, 36, 37, 44, 53, 59, 80, 81
Italy, 75

J

Japan, 7, 22, 23, 75, 89, 97, 98
joint ventures, 98
juice processing, viii, 87, 88, 92, 93, 96, 97, 105

K

Kenya, 75, 76
kidney, 40, 41, 50, 73, 76, 77
Korea, 75, 76, 133, 136
Kyrgyzstan, 74

L

labor force, 123
land abandonment, 32

land tenure, 33
large-scale farms, vii, 39
Latin America, 5, 30, 34, 54, 82, 84
laws, 2, 5
lead, 65, 103, 108
liberalization, 43, 45, 120, 122
lifestyle changes, 123, 124, 130
light, 40, 42, 59
liquidity, 62
livestock, 56, 59, 120, 121
loans, 6, 60, 64, 96, 101, 102, 103, 104
local government, 102, 103, 107, 109
logistics, 29, 30
low production costs, vii, 1
low temperatures, 8
lower prices, 24, 28, 93
low-interest loans, 104

M

machinery, 62
majority, 24, 29
Malaysia, 26, 75, 121
management, 2, 8, 28, 31, 32, 33, 36, 60, 106, 107
marginal product, 108
market share, 20, 24, 26, 27, 32
marketing, 12, 25, 43, 44, 45, 46, 47, 48, 54, 57, 60, 64, 65, 78, 83, 84, 98
materials, 97, 109
matter, 14
measurement, 132
meat, 70, 123, 129
media, 112
median, 56, 57
Mediterranean, 23
membership, 23
merchandise, 3, 4
metals, 2
Mexico, vii, viii, 14, 15, 18, 20, 23, 24, 25, 26, 29, 39, 40, 41, 42, 43, 44, 45, 46, 47, 48, 49, 50, 52, 53, 54, 56, 57, 58, 59, 60, 61, 63, 67, 68, 70, 71, 72, 74, 75, 76, 78, 79, 80, 82, 83, 84, 92
middle class, 118
migration, 122
moderates, 8
modifications, 59
molds, 106
monopoly, 119
Morocco, 7
mortality, 127
mortality rate, 127
multinational companies, 96
multinational corporations, 15
mung bean, 65
Myanmar, 72, 74, 76

N

NAP, 65
national income, 3
natural advantages, 2
negative consequences, 106
Netherlands, 26, 75
New Zealand, 26, 90
Nicaragua, 74, 76
Noninsured Crop Disaster Assistance Program, 65
North America, vii, 18, 39, 40, 44, 71, 74, 82, 85, 105
North American Free Trade Agreement (NAFTA), vii, 18, 39, 40, 41, 42, 43, 45, 46, 53, 59, 71, 73, 74, 75, 76, 83, 84
Northeast Asia, 5, 22, 23, 33
nutrients, viii, 39, 40, 67, 72, 127
nutrition, 127

O

obstacles, 7, 31
officials, viii, 87, 88, 92, 98, 102, 104, 105, 106, 107, 108, 109
oil, 124, 129
open markets, 128
operations, 15, 97
opportunities, viii, 19, 58, 72, 83, 117, 118
ores, 2
organize, 62

outsourcing, 15
overlap, 10
overproduction, 28
ownership, 6, 97

P

Pacific, 3, 8, 30, 31, 50, 136
Pakistan, 75, 76
palm oil, 119, 121, 129
participants, 62, 63, 64, 65
pasteurization, 106
per capita expenditure, 67
Peru, v, vii, 1, 2, 3, 4, 5, 6, 7, 8, 9, 10, 11, 12, 13, 14, 15, 18, 19, 20, 21, 22, 23, 24, 25, 26, 27, 28, 29, 30, 31, 32, 33, 34, 35, 36, 74
pesticide, 13, 106, 107
pests, 28
petroleum, 3
Philippines, 26
plants, 11, 15, 88, 95, 97, 102, 104, 106, 114
platform, 134
Poland, 89
police, 107
policy, 2, 40, 118, 119, 121
policy reform, 121
population, 30, 31, 69, 70, 85, 122, 124, 127, 130, 133
population growth, 124
portfolio, 2, 28
Portugal, 75
poultry, 70, 129, 130, 134
poverty, 62, 102, 118, 134
poverty alleviation, 102
precipitation, 42
predictability, 8
preferential trading arrangements, vii, 1
preferential treatment, 59
president, 32, 98
price effect, 60
price signals, 56
private investment, 31, 101
privatization, 6, 35
processed artichokes, vii, 1, 2, 3, 10, 20, 21, 33
producers, vii, 2, 4, 6, 10, 12, 13, 15, 18, 19, 20, 21, 23, 31, 36, 39, 40, 42, 43, 45, 46, 49, 54, 57, 58, 60, 61, 62, 63, 64, 65, 66, 70, 71, 76, 83, 84, 96, 118, 119
production costs, vii, 1, 8, 25, 92
productivity growth, viii, 117, 118, 120, 122, 134
professionals, 62
profitability, 15
project, 31, 32, 59, 61, 62, 108
property rights, 6, 32
proteins, 67
public-private partnerships, 31
purchasing power, 123, 134
purchasing power parity, 123

Q

quality control, 29
quality standards, 21
quotas, 40, 71, 83

R

rainfall, 8, 98
rape, 27
rate of return, 120
raw materials, 88, 98, 100, 105, 106, 108, 109, 114
real estate, 102
recall, 106
recommendations, 70
recovery, 62, 129
recruiting, 88
redistribution, 31
reform, 31, 94, 111, 135
reforms, 6, 31, 121
relief, 18
remediation, 107
representativeness, 64
requirements, 123
researchers, 83

residues, 106, 107
resources, 61, 101, 119
response, 45, 56, 118, 129
restaurants, 125
restrictions, 32, 41, 65, 71
restructuring, 71
retail, 84, 109, 124, 125
retirement, 54
revenue, 65
rights, vii, 1, 2, 6, 28, 31, 32, 119
risk, 14, 33, 60, 62
risk management, 60
risks, 60, 106
root, 124
rubber, 119, 129
rules, 60
rural areas, 60, 123
rural poverty, 31
Russia, 7, 26, 118
Rwanda, 76

S

safety, 102, 105, 106, 107
savannah, 49, 50
scarcity, 3, 98
science, 83
scope, 48
seafood, 124
security, 30, 118, 126, 127, 134
seed, 42, 50, 73, 82, 118
self-sufficiency, viii, 117, 120
sensitivity, 43
services, 7, 29, 60
sewage, 102
sheep, 84
shelf life, 13, 14, 29
signs, 56
Singapore, 26, 97
social justice, 31
society, 40
soil type, 98
South Africa, 7, 22, 75, 90
South America, 3
South Korea, 7, 22, 23, 98
soybeans, 58, 59, 119, 120, 130, 133
Spain, 12, 18, 20, 26, 75
specialty crop, 78
species, 41, 83
Spring, 55, 63
stability, 119
state, 6, 11, 78, 98
statistics, 40, 41, 42, 43, 46, 49, 52, 53, 58, 59, 63, 64, 66, 73, 74, 83, 84, 91, 94, 95, 99, 101, 103, 106, 111, 136
stimulus, 5
stock, 96
storage, 29, 54, 60, 61, 104
structural reforms, 6
structure, vii, 39
subgroups, 70
subsistence, 54
sulfur, 42
supplier, 2, 4, 10, 15, 18, 43, 50, 88, 108
suppliers, viii, 19, 23, 24, 29, 89, 108, 117
supply chain, 107
surplus, 60
sweeteners, 124
swelling, 130

T

table grapes, vii, 1, 2, 3, 10, 21, 22
Taiwan, 22, 26
takeover, 32
Tanzania, 76
target, 11, 77, 85, 108, 127
tariff, 18, 40, 41, 73, 74, 83, 128
tax breaks, 101, 102, 103
technical assistance, 62, 64
technical change, 119
techniques, 8
technologies, 102, 118, 119
technology, 6, 11, 64, 120, 121
technology transfer, 121
temperature, 8, 98
tension, 32
testing, 106, 107, 108
Thailand, 26, 74
total factor productivity, 119, 120

total product, 121
trade, vii, viii, 2, 3, 5, 6, 7, 9, 13, 20, 29, 32, 36, 39, 40, 41, 42, 43, 44, 45, 57, 71, 74, 83, 84, 89, 90, 98, 104, 110, 117, 118, 119, 121, 127, 128, 130, 134
trade agreement, 2, 5, 6, 32, 84
trade liberalization, viii, 117, 118
trade preference, 9
trading partners, 5, 7, 9, 28
trafficking, 6
trajectory, vii, 1
transformation, 95, 118, 130, 134
transformations, 133
translation, 114
transportation, 7, 14, 19, 28, 54, 123
transportation infrastructure, 19
treatment, 20, 36, 102
trial, 19
tropical perennial crops, viii, 117, 119, 120, 129
Turkey, 75, 133, 136
turnover, 7

U

U.S. Department of Agriculture (USDA), 9, 13, 14, 15, 17, 22, 23, 24, 31, 40, 41, 44, 45, 46, 47, 49, 50, 52, 53, 55, 59, 66, 70, 71, 73, 74, 75, 76, 77, 83, 84, 85, 90, 91, 93, 94, 95, 96, 99, 100, 101, 103, 106, 113, 119, 128, 129, 136
U.S. Department of Commerce, 44, 45, 74, 114
U.S. Department of Labor, 47, 82
unemployment rate, 7
unions, 30
United Kingdom, 25, 26, 33, 74, 75
United Nations (UN), 2, 34, 35, 75, 76, 78, 84, 119, 129, 131, 134, 136
United States, vii, viii, 1, 2, 3, 4, 5, 6, 10, 11, 12, 13, 18, 20, 21, 22, 23, 24, 25, 26, 27, 28, 30, 33, 35, 39, 40, 41, 42, 43, 44, 45, 46, 47, 48, 49, 50, 52, 53, 54, 56, 57, 65, 67, 70, 71, 72, 73, 74, 75, 76, 81, 82, 83, 84, 87, 88, 89, 90, 91, 92, 107, 114, 117, 125, 132
universities, 106
urban, 48, 67, 119, 122, 123, 125, 130
urban areas, 67, 119, 122, 123
urban life, 123
urban population, 122
urbanization, 123, 124

V

value added tax, 104
varieties, 20, 22, 24, 26, 40, 41, 42, 43, 50, 57, 58, 61, 83, 102, 105
vegetable exporter, vii, 1, 2, 8, 28, 30
vegetable oil, 114, 124
vegetables, vii, 2, 4, 5, 8, 18, 29, 35, 65, 67, 70, 123, 124, 132
Venezuela, 75
Vietnam, 26
volatility, 60
vulnerability, 28, 56, 108

W

wages, 7, 122
Washington, 2, 14, 15, 34, 36, 135
water, vii, 1, 2, 3, 8, 11, 28, 30, 31, 32, 33, 36, 49, 50, 61
water quality, 37
water resources, 31, 32, 37
water rights, 31
water supplies, 30, 31
watershed, 30, 31
weakness, 105
wealth, 123
weather patterns, 8
web, 102, 110
welfare, 102
wells, 8
wholesale, 47, 48, 54, 84, 104, 105, 106
workers, 7, 30, 35
working hours, 7

World Bank, 3, 4, 6, 31, 34, 36, 37, 54, 82, 84, 97, 119, 120, 122, 123, 124, 125, 131, 133, 136
World Trade Organization (WTO), 83, 128
World War I, 120
worldwide, 22, 36

Y

Yale University, 134
yield, 36, 51, 64, 65, 71, 105, 106
yuan, 93, 98, 104, 109, 114